Beginner's Guide to
Heating and Insulation

Beginner's Guides are available on the following subjects:

Amateur Radio
Audio
BASIC Programming
Building Construction
Cameras
Central Heating
Colour Television
Computers
Digital Electronics
Domestic Plumbing
Electric Wiring
Electronics
Fabric Dyeing & Printing
Gemmology
Home Energy Saving
Integrated Circuits
Making Wooden Furniture
Microprocessors
Minerals
Photography
Processing & Printing
Radio
Spinning
Super 8 Film Making
Tape Recording
Technical Illustration
Technical Writing
Television
Transistors
Video
Videocassette Recorders
Weaving
Woodturning
Woodworking

Beginner's Guide to
Heating and Insulation

W. H. Johnson

C.Eng., F.I.GasE., F.I.S.T.C.

Newnes Technical Books

Newnes Technical Books

is an imprint of the Butterworth Group
which has principal offices in
London, Boston, Durban, Singapore, Sydney, Toronto, Wellington

First published 1984

British Library Cataloguing in Publication Data

Johnson, W. H. (William Harold), 1913-
  Beginner's guide to heating and insulation
  1. Dwellings–Heating and ventilation–Amateur's
  manuals   2. Dwellings–Insulation–Amateur's manuals
  I. Title
  697   TH7224

  ISBN 0-408-01362-1

Library of Congress Cataloging in Publication Data

Johnson, W. H. (William Harold), 1913-
  Beginner's guide to heating and insulation.

  Includes index.
  1. Heating.   2. Insulation (Heat)   I. Title.
  TH7223.J58   1984   693.8'32   83-17429
  ISBN 0-408-01362-1

Photoset by Butterworths Litho Preparation Department
Printed in England by Butler & Tanner Ltd, Frome, Somerset

Preface

This book is aimed primarily at householders who want a broad view over the twin subjects of heating and insulation, usually with a view to doing something about at least one of them. This need not mean practical involvement. The buyer who is able to discuss technicalities with a contractor is in a strong position compared with one who is not. Students in their elementary year might gain something from taking a broad view, and in particular relating insulation to heating, and solo to central heating. This is a developing phase which the standard books have not yet been adapted to cover. Perhaps even some members of the trade might get an idea or two. This could be the case with someone who has spent too many years making a living, with no time to take a leisurely look around at trends and fashions.

As for those who are enthusiastic about DIY insulation and heating, this book aims to help them to make as good a job as possible. But there must be limits, determined by professional skill professionally acquired, often backed up by apparatus not likely to be owned by someone doing one job. When that limit is reached, the book states quite flatly that now is the time to hand over to an expert. This applies in particular to commissioning gas- and oil-fired appliances, where mistakes could have more than economic consequences.

It is a pleasure to record our thanks to those manufacturers and trade associations who so willingly gave illustrations and information. They are acknowledged individually in the appropriate places in the book.

W.H.J.

Contents

1

What insulation is about

The main purpose of insulation is to try and prevent heat escaping from the inside of a house to the outside. Although insulation cannot stop heat loss completely, it can reduce it considerably.

It is entirely in accordance with the contemporary energy equation that insulation shall come before heating. For too long people have been deciding to have heating put in, and then next year, or the year after, or some time, they will think about insulation. Nowadays, if you have a certain limited amount of money to spend, it is best to spend it on insulation rather than on heating since the better the insulation, the less important the heating.

The economic case for insulation is strong. The cost of insulation rises very little, whereas the cost of fuel rises all the time. Heat lost costs the same as heat used, and as each unit lost increases in cost the payback time for money spent on insulation grows shorter, and insulation more and more attractive.

Figure 1.1 gives an idea of how heat is lost, and also what savings can be achieved. In a fairly average construction such as is shown, the payback time on loft insulation can be measured in months, usually less than two years. A major item such as cavity-wall insulation may be reckoned to pay back within five years and is a good proposition. Insulation has the disadvantage that its savings are not always spectacular and are often hidden, and it does not give the immediate and positive result that a new boiler gives. But insulation is a

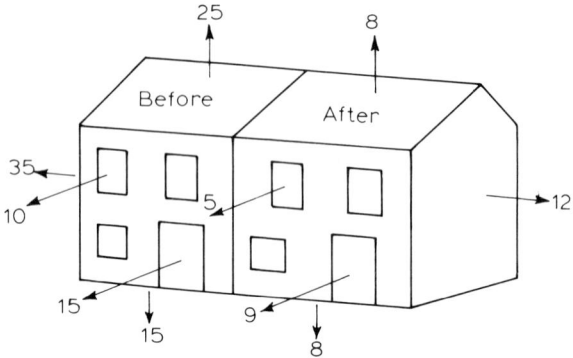

Figure 1.1. Heat loss before (left) and after (right) insulation (typical results). The left-hand house losses total 100. At the right the losses total 42. Heat loss is therefore more than halved

servant with neither running nor maintenance cost and broadly speaking is only as good as its installation. This is a challenge to the person installing, to frustrate the natural tendency of warmth to escape to somewhere colder.

Although a huge number of materials can be considered as insulators, only a few are suitable in practice; nowadays, these are preponderantly synthetic. This is explained by the example of the thatched roof. A good thatch is more effective than even a fully insulated conventional roof. However, thatch can harbour rodents, birds and insects – in short it supports vermin. Straw is in the same category but has been rescued by slabbing it with cement for use as rigid insulation. Most organic materials offer a sympathetic base for vermin, moulds and fungi, which synthetics do not. Glass fibre, for instance, would make a very uncomfortable lodging. Hygroscopy, a tendency to attract water vapour from the air, is another disqualification, for wet materials do not insulate.

It would be a mistake to think of all insulators as solid and substantial. Liquids can act as insulators, though rarely in domestic applications. (Industrially, very cold conditions can be maintained by having a surrounding vessel with liquid CO_2 or nitrogen in it.)

Gases are more common than liquids as domestic insulators. The commonest gas is air, which is present in double glazing. Sealed double glazing consists of two panes sealed in a frame separated by a gap of 18–20 mm (*Figure 1.2*). The

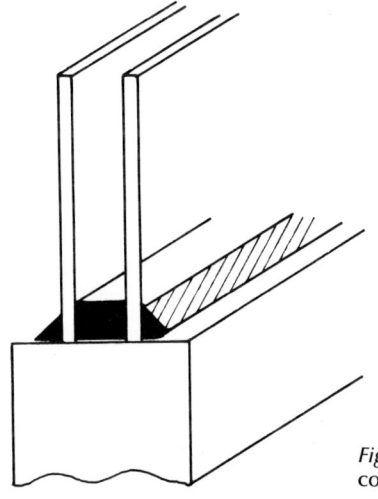

Figure 1.2. Double glazing is a composite insulator

gap is filled with *dry* air, for if it is wet it will act as a conductor. If the gap is too narrow it will lose insulating value. If the gap is too wide, as it often is in 'secondary glazing' (i.e. where a second pane is added to the original), the result can be seen in *Figure 1.3*.

The extra width allows for two streams of air to pass, one rising and one falling. Warmth from the room causes the inner stream to rise, carrying that warmth, until it turns over at the top and comes down, giving up its warmth to the cold outer pane of glass. The cooling assists the circulation, of course, and the whole process succeeds in transferring warmth from indoors to outdoors through the glass. That, you will remember, is what it was specifically intended *not* to do.

Air is the principal ingredient in a number of insulating materials, such as expanded foamed plastics, polystyrene

3

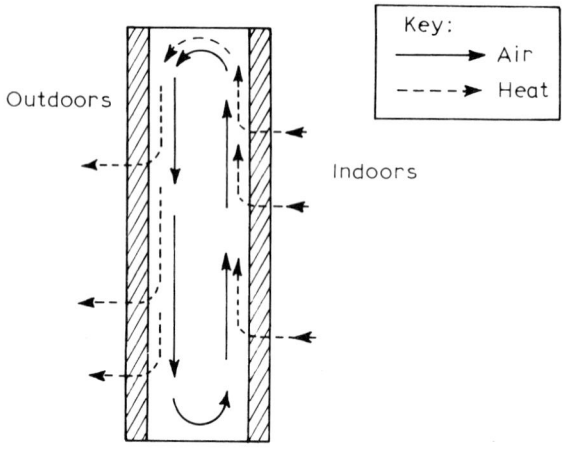

Figure 1.3. Too wide an air gap allows internal circulation of air, which transfers warmth from the indoor side to the outdoor side

and polyurethane, and glass and mineral fibres which are mainly a mechanical means of trapping the air in small pockets, or in a 'jungle' which immobilises it.

But still there are plenty of solid materials which are insulators because they are bad conductors of heat, and they may be used by themselves or in combination. Asbestos (which is not dangerous in ordinary applications) is a long-standing example. So are magnesia and compounded materials like plasterboard.

In every case the effectiveness of an insulator is proportional to its thickness measured in the direction of travel of the heat. The same rule applies to materials not usually considered to be insulators. A stone wall of (say) 150 mm would be reckoned a very bad thermal barrier. But a stone wall a metre thick is a prime factor in many a snug old cottage. Thickness is the principal reason why there is a move to make the gap in cavity walls 100 mm instead of 50 mm, to accommodate more insulation. Left unfilled it would suffer from the trouble illustrated in *Figure 1.3*, internal circulation of air.

Reflective insulators

One class of insulator in which thickness plays no part is that known as 'reflective'. A few years ago aluminium foil was hailed as a straight equivalent of the bulk insulators, the mineral fibres and so on, in particular for loft insulation. More recently it has found its best uses in industrial work, often compounded with more rigid backing, where the environmental problems differ from the domestic ones. There is still a good domestic use for reflective foil, but in future it must be used with discrimination. For foil is also a first-class vapour barrier, and we shall be seeing in a later chapter the danger of preventing the free passage of water vapour away from structural timber. To see how this works, the alternative to laying glass fibre etc. between joints in lofts was to spread foil over the joists. We now know that doing this allows water vapour to accumulate on the joists, leading to dry rot. The correct and logical way to use foil is therefore at the front end of the process, and at best used to 'paper' the ceiling beneath. This arrests water vapour *before* it can get into the timbers.

U-value

To evaluate and compare insulating materials, the most useful simple guide is the U-value. This is the amount of heat which will pass from the atmosphere on one side of the material to the atmosphere on the other side, the thickness of the material being as supplied, and considering unit area for unit time and for unit difference of temperature. In less severe language, more heat will pass through a bigger area or over a longer time or if there is a bigger difference of temperature.

All U-values are now given in metric units, $W/m^2 \, °C$, but if you should come across imperial units, $Btu/h \, ft^2 \, °F$, and wish to compare values, you must multiply the imperial figure by 5.7 to get the metric equivalent.

All that you need to know is that a high U-value is worse for you than a low one. Armed with that simple fact you can

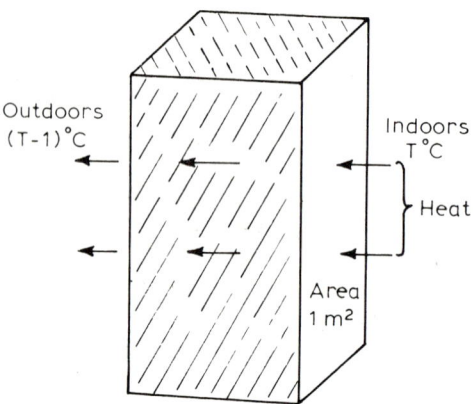

Outdoors
(T-1)°C

Indoors
T°C

Heat

Area
1 m²

Figure 1.4. Conditions for measuring U-value

usefully decide whether a merchant is offering you, or an installer intending to fit, the insulator with the best insulating value. Be prepared though to be told that there might be other, e.g. structural, reasons for the choice. Your local library should be able to get you the annual Insulation Handbook, which gives a comprehensive list of proprietary insulating materials and their properties including U-value.

Just two cautions about reading the Handbook, though. Quite a large number of insulating materials are marked 'high temperature'. They are not suited to, and do not insulate very well at, domestic temperatures. Many others are made with particular reference to industry and would not transplant at all well into a house. Perhaps the best use of the reference book is to make your own check on what you are being offered.

If you read a book on insulation, particularly the Handbook, you may be struck by the enormous variety of materials on offer, some of them chemically the same but under different brand names. There is nothing against using less familiar ones, so long as U-value, physical properties, price and availability are satisfactory. We can do no more here than glance at the most common and offer a few comments on each.

• *Cavity walls.* Existing walls must be done by a contractor, who will use foam, blown mineral fibre or plastic pellets. There is little to choose so long as the contractor is reputable, preferably a member of NCIA. But for new work, for instance if you are adding an extension, the cavities can be insulated as they are built, using rigid resin-bonded glass-fibre boards. Detailed instructions can be obtained from Eurisol UK.

• *Attic floors, and false walls behind cladding boards.* Non-rigid materials are suitable for this application. There is the pour-in vermiculite; this is easy to handle, non-irritant, and easy to level off. However, it is not so easy to contain at joist ends, and is easily disturbed by draughts, which are nowadays recognised as important in attics. In the same category are the plastic pellets mentioned above for cavities. Sheet polyurethane and polystyrene, of about 25 mm thickness, is harmless to handle. It can be cut to be an interference fit inside joists, using a sharp knife or preferably a hacksaw. It tends to crumble at cut edges and for that reason can be made a tight fit using a flexible material, such as mineral fibre. The favourites are glass fibre and mineral fibre, which should be handled wearing gauntlet gloves but otherwise have everything in their favour. They are supplied in ready-made thicknesses, 50 or 80 or 100 mm, to standard widths and in rolls of stated length. To cut off required lengths, either tear carefully (wearing gloves), cut with heavy cutters such as shears, or tease apart using a sharp knife.

Water pipes should be either buried well down in the insulation, for maximum protection, or taken above it and given their own protective wrapping. Electric cables should *always* be run outside the insulation, partly because they need to lose warmth since they could become overheated, and partly because some modern insulating materials can attack pvc coverings.

• *Roof pitch insulation.* The materials suited to this job are expanded polystyrene and polyurethane, and glass or mineral fibre. It is taken for granted that all such insulation will be retained by cladding sheets tacked to the underside of the roof timbers. Both materials have been discussed above.

• *Underfloor insulation.* The nature of this ground-floor

application is, as we shall see, such as to require some rigidity in the insulator, since solid cladding sheets cannot be used. In consequence we must look to the expanded plastics, and to resin-bonded sheets of glass and mineral fibre. The need for tightly fitting insulation is as already described.

● *Thermal inner linings.* These are means of insulating walls from the indoor side. We have already seen something of these in the section on attic floors. Another method, handier though less effective, is to use expanded polyurethane or polystyrene sheet or tiles, stuck on with a special plastics tile adhesive. Unless you buy sheet in widths which can be used without much cutting, a neater and more coherent job can be made with tiles. So long as the wall surface will not become wet, i.e. it is an inner cavity or has been waterproof painted, then a blob of adhesive at each corner and one in the middle is all that is needed. Press each tile firmly into place. And since the surface is crushable by furniture etc, the protection of a sheet of hardboard or similar might sometimes be desirable.

Note: these plastics tiles are very useful for ceilings in such conditions as when the attic space is inaccessible. It is always best to ask for fire-retardant grades of tile or sheet.

It seems ironic that even today one feels impelled to put in a good word for insulation, to have to 'sell' the idea. After all the talk, energy conservation propaganda and 'Save It' stamps, it does not spring to mind as heating does. Perhaps that is because it is mainly invisible and unexciting, and usually needs a year or so's fuel bills to prove its value. But we ought to be able to make an exception for walls and windows, where we would expect an immediate increase in comfort. The reason lies in the fact that cold walls and windows give cold radiation. And just as radiant warmth is more potent and more effective than the warmth of warmed air (see Chapter 12), so the same is true of cold radiation versus cold air. Cold walls and windows are very chilling and depressing.

Insulation, by slowing down heat loss, speeds up the warming of the surfaces by the heating, and so lessens the discomfort of cold radiation. That is, in human terms, a bonus, but insulation tends to be subject to, if not dominated by, the concept of cost-effectiveness. This is easily explained. Suppose that you have a job done, say cavity walls filled, at a cost of £x, and which will save you £y a year in fuel bills. By dividing x by y you will get a payback period, x/y, which is years and is a measure of cost-effectiveness. In the example it might be five years, a fair average for the type of job. But as fuel prices rise, the annual saving, hence the payback period, changes for the better.

Estimates for payback time vary (though not widely among competent observers) and sites are not all alike. At a fair consensus you could find loft insulation quoted at one year, double glazing (installed) at 25 years, on the face of it quite absurd. But double glazing is still popular, a salutory reminder that life is not run by economists and accountants. What price is put on comfort, for instance?

Some typical U-values

	W/m² °C
225 mm solid wall, plastered	1.9
Ditto with 25 mm inner lining	0.8
275 mm cavity wall plastered	1.5
Ditto cavity foam filled	0.5
Brick/insulating block cavity wall plastered	1.0
Ditto foam filled	0.4
Roof unsarked, slates	to 3.0
Ditto with 100 mm mineral fibre in loft	0.3
Ditto with 160 mm (recommended thickness)	0.2
Windows: single-glazed according to exposure	4.5 to 5.0
Double-glazed to high standard (20 mm gap)	2.8
Floor: suspended 20 mm board bare	0.8
Ditto with 100 mm insulation	0.3
Solid concrete	0.8
Ditto with 75 mm insulation	0.3

Note: none of the above figures is absolute, since many local conditions can introduce some variation. But such data are indicative, and they show the very large percentage savings to be expected from good-quality insulation properly applied.

2

Insulating the top of the house

In general this is the easiest job in the insulation schedule, and, as *Figure 1.1* shows, this is the place where a quarter of all heat loss occurs. It is also generally the best investment, in terms of fuel savings for cost of installation. And then, probably because of those virtues, it is the one for which the Government is prepared to give a grant. All you need to qualify is to have a totally uninsulated loft, and you can get details at your local town hall.

But here are a few of the important points.

1. The elderly and disabled on low incomes can qualify for a 90 per cent grant, the maximum now being £95.
2. The standard rate of grant is 66 per cent, and the maximum is now £69.
3. The depth of mineral-wool insulation is now 100 mm (4 in), and the local authority will supply a list of approved substances from which you must choose. You buy first and claim repayment afterwards.
4. Insulating the cold water cistern and associated pipes, and also the hot water cylinder, are conditions which must be observed.
5. The prices given in 1 and 2 above relate to material, and people of average ability should be able to convince the town hall that they can do the job. If using a contractor, his installation fee can be added in for grant purposes.
6. If there is no access door to the loft, this too may be a grant-aided item.

7. You must not start anything until the town hall says so.

8. If you want a contractor but do not know where to choose, write to the National Association of Loft Insulation Contractors for a list of local members. The address is given in Chapter 14.

Earlier we stressed the *general* situation, and this is the one where the floor of the attic is insulated. But we shall also be looking at the other case, where the attic is incorporated in the living space, and the insulation must therefore go on the outside of the attic. Both cases pose their special problems, and we will start with the floor insulation.

Materials

The material most used is mineral-wool fibre, which is melted and blown rock. Glass fibre is very similar. Quite different is vermiculite, which is lightweight flakes of mica bearing some resemblance to fish scales.

The Building Regulations, which control these matters and set the conditions for a grant, require a minimum of 100 mm (4 in) of blanket, or 150 mm (6 in) of vermiculite. As fuel prices rise there is a growing case for exceeding that minimum, and at present there is a strong case for 150 mm of blanket, even though the extra would not rank for a grant. That is a decision for the reader to make. Blanket is sold in rolls of stated thickness, and may be cut to size with a sharp knife or saw. Gloves, preferably gauntlet type, should be worn. A face mask is recommended for anyone who is affected by dust.

There is rarely any advantage in looking beyond the materials named, since they offer the best specification at the best prices. But there are other materials, often proprietary ones, and if they appear on the local-authority approved list they may be taken to be satisfactory.

The method
Forget about stretching the blanket across the joists. What seems simple is very difficult to finish off successfully, and it

is far simpler to make use of the channels formed by the ceiling joists.

Measure these channels, their number, length and width between the joists. The first two figures will give you the length of blanket needed, and this should be, or be cut to, approximately 50 mm (2 in) wider than the channel width. This is so that the blanket will fit snugly into the channels with no gaps (*Figure 2.1*). Keep electric cables outside the insulation.

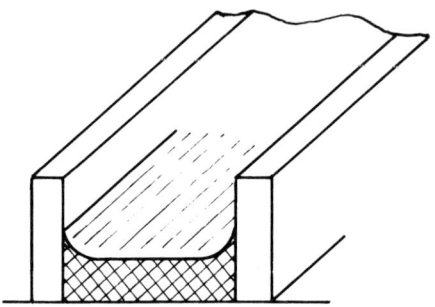

Figure 2.1. Mineral-wool blanket should be cut to a width of about 50 mm more than the channel between the joists

The next job is essential if using vermiculite, and desirable even with blanket. This is the business of placing end-stops near the end of each joist channel, so that the insulating material cannot creep into the eaves. The reason for this will appear shortly. These end-stops may be nothing more than a brick on its side, a block of wood, or hardboard cut to size and wedged between the joists (*Figure 2.2*).

Work from two planks balanced across the joists, so that you may move one while kneeling on the other. It helps to pre-measure your rolls of material, so that by taking up a mid-position you may push a roll to unroll it and it will travel as far as the end-stop. It is then not difficult to make a good join in the middle. If you can do this, you can also make good use of a simple tool, shown in *Figure 2.3*, which is a piece of hardboard or light wood nailed to the end of a pole. With its

12

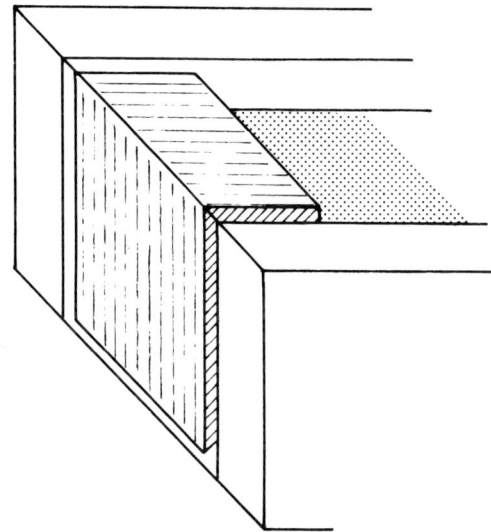

Figure 2.2. An end-stop at the end of each joist channel stops insulating material from creeping into the eaves

help you can lightly tamp down the roll into the channel, all the way, from your position in the middle of the floor.

If you use vermiculite you will need instead a homemade device, similar to the one in *Figure 2.3*, which acts as a rake to level the material to the desired depth. Vermiculite should not be used in an attic in which there are severe draughts, since these would probably shift the material.

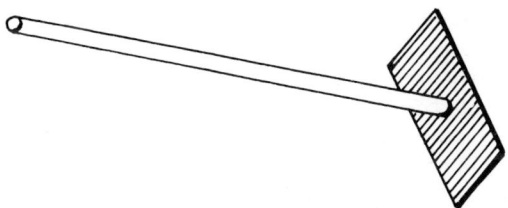

Figure 2.3. Device to help fit roll of insulating material between joists

Preventing damp

With the insulation complete we can look into the next essentials, which are a consequence of the insulation. Previously, warmth from the room below travelled through the ceiling, carrying water vapour, and the warmth dispersed this into the general atmosphere, harmlessly. Now, the water vapour travels through the ceiling and through the insulation, but the warmth does not. Consequently the vapour condenses as soon as it becomes cool, mainly in the insulating material, and on the ceiling timbers, joists and so on where it can cause dry rot. Obviously this is most unsatisfactory, but steps can be taken to deal with it.

1. *Never* lay a vapour barrier, such as polythene sheet, over the insulation.
2. Be very careful about flooring the attic, for example if it is used as a storage space. Try to floor only a part or parts, and even then avoid tongued and grooved timbers. And do not lay floor boards tightly together. In short, give the insulation a chance to 'breathe'. Never put down carpet, lino or other floor coverings.
3. Then, the positive action, take steps to encourage through-draughts. This is very different from former practice, but essential. The modern attic, and the modern garage, ought to be cold but dry, with ample ventilation.

For most roofs, those whose pitch is greater than about 15°, all that is needed are draught openings in the soffit board or eaves at opposite sides of the house (*Figure 2.4*). If openings have to be made for the purpose, these should be the equivalent of a 10 mm slot along the whole length, usually in practice being holes drilled in the soffit. For lesser pitches, the slot equivalent rises to 25 mm and in addition a high-level air outlet is needed. This may be an air brick or bricks in the gable end(s). Or it may be a ridge-tile ventilator, both of which are likely to be builder's work. See *Figure 2.5*.

Let us look now at the cases where the attic is a part of the living space of the house and has therefore to be kept reasonably warm. Clearly the insulation must go on the

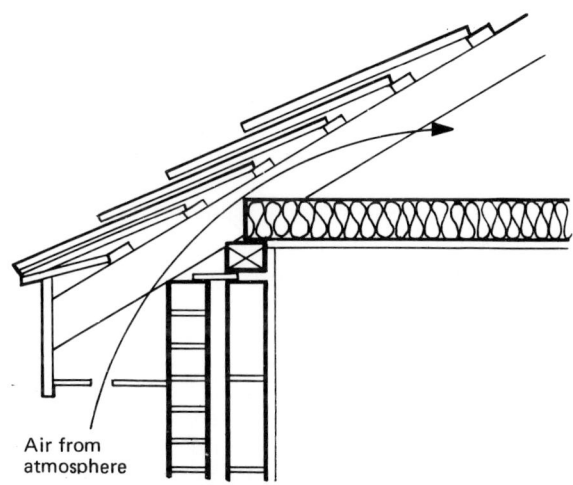

Air from
atmosphere

Figure 2.4.   Where a roof has a pitch greater than 15°,
draught openings should be made in the soffit, usually by
drilling

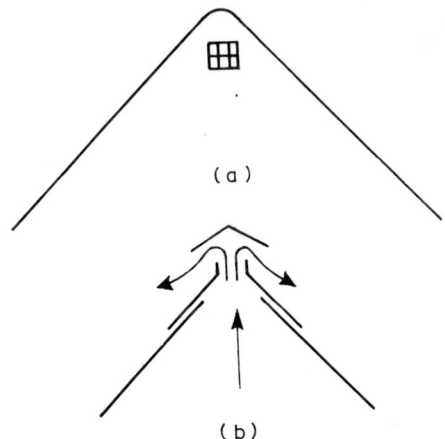

(a)

(b)

Figure 2.5.   For pitches under 15°, a
high-level air outlet is needed, in the form of
either (a) an airbrick in the gable end or (b) a
ridge-tile ventilator

outside of the attic, and follow the pitch of the roof. Very often roof spaces which have been used for some time already have ceiling board fitted. Hardboard is commonly used, with wallpaper covering. In order to insulate properly this must be removed, and if done with care most of the sheets can be reused.

The roof proper usually consists of slates or tiles direct on to the roof timbers, or roofing felt is interposed. For insulating purposes the difference is mainly one of degree. Without any felt, you can be sure that some wind will whistle in between the slates. The possibility that some rain may follow it cannot be overlooked. With felt present and in good condition the wind factor will be much less, but rarely absent. Any entering rain will be kept behind the felt. The essential difference lies in the absolute rule that insulation must stay dry, because if wet it does not insulate. And so, if there is no roofing felt, waterproof insulating material is needed.

We now have to reiterate the need for caution shown in the case of ceilings, principally the importance of ventilation in dispersing moisture once warmth is absent. Here are three rules of utmost importance, all aimed at preventing timber decay.

1. Any structural or other timber which is to be hidden away must be given three generous coats of an anti-woodworm anti-dry-rot fluid, with soaking-in time between each.

2. On the *warm* side of the insulation have a vapour barrier, a common material being heavy-gauge polythene sheet, joins in the sheets being double-folded and stapled, with scotch tape over the staples.

3. On the *cold* side of the insulation, leave space for a current of air. As indicated, feltless roofs and most felted ones are not short of air currents adjacent to the underside of the roof. But if you should happen to have a roof without any indication of draught, stop the insulation short by say 50 mm, at top and bottom ends so that a current of air may flow up behind it. (This is what you do with disused blocked-off chimneys.)

The roof

Coming to do the job we start with the roof proper, and here are the steps to take.
1. Check the condition of the roof and have any defects put right.
2. Apply the preservative fluid.
3. Cut strips of insulation material to size. Unlike the ceiling job, these should be a good holding fit but without a tuck-in allowance.

a) for roofs with felt, assumed dry, ordinary mineral- or glass-wool blanket is satisfactory.

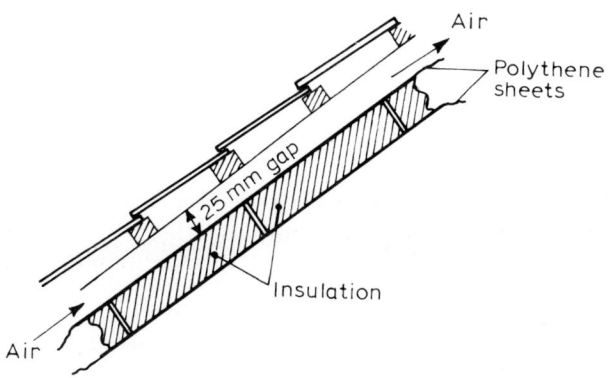

Figure 2.6.   Roof with no sarking

b) If moisture could be present, use either resin-bonded glass fibre or expanded polystyrene or a proprietary material with the right qualities.
4. Fit the insulating strips into the joist spaces so that they are flush with the indoor edge of the joists and allow an air space behind. For example, if the joists are 150 mm (6 in) deep and you use 100 mm (4 in) insulation you can count on having a 50 mm (2 in) gap behind.
5. Next, fasten the vapour barrier to the indoor side of the joists. Drawing pins will do, since the cladding sheets will eventually give all the support needed.

6. Finally, nail or screw on the cladding sheets, which may be hardboard, ceiling board, or any similar material. If using a thicker board than hardboard, usually one which has a plaster sandwich, the joins can be filled with a plaster substitute such as Sirapite or Polyfilla.

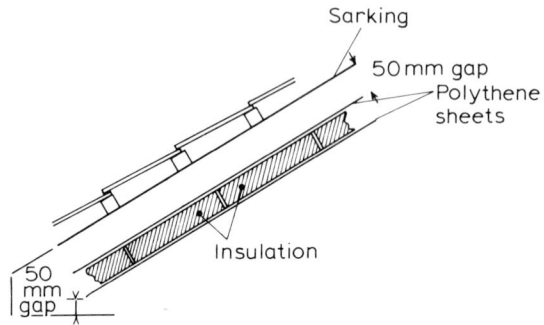

Figure 2.7. Roof with sarking

7. Instead of steps 5 and 6, you can combine them by using a foil-backed board, i.e. one which incorporates its own vapour barrier.

Note: screwed on sheets are preferred to nailed-on ones, partly because nailing can disturb slates, and partly because removal at a later date is so much easier. A little stiff grease applied to screw threads might help to avoid seizure.

8. The cladding may be painted or papered as desired. Or to combine decoration with extra insulation, you can stick expanded polystyrene tiles, of fire-retardant quality, on to the cladding face. These are in themselves presentable, or they may be painted with an emulsion paint.

Flat roof

The need for great care, which is apparent in all roof insulation, is at its greatest when dealing with flat roofs. These are by definition any roof from horizontal to 10° inclination, sometimes decked in concrete, more often felt on timber. Felt is not a permanent material and can let rain through by tearing or rotting. The first step before insulating

18

is not only to make sure that the roof covering is of good-quality material in first-class condition, but also to satisfy yourself that no deterioration has occurred, mainly in the timber, from any previous weakness in the covering. Another important step is to treat all the wood you can reach with three good coats of an anti-woodworm and anti-dry-rot liquid.

There are two methods of dealing with flat roofs, known as the 'cold roof' and 'warm roof' methods. The second is not much used domestically and for that reason we show no more than a sketch to indicate what is involved (*Figure 2.8*).

As *Figure 2.9* shows, the principles applied to pitch insulation apply here. The vapour check is on the warm side of the

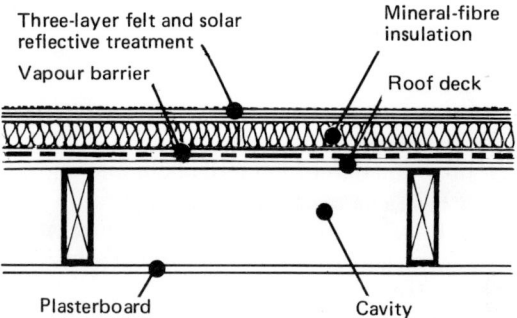

Figure 2.8. Warm-roof construction (courtesy Eurisol UK)

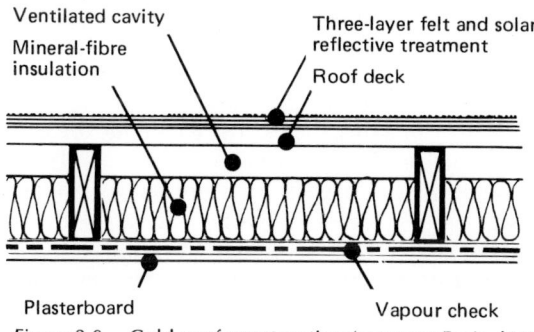

Figure 2.9. Cold-roof construction (courtesy Eurisol UK)

19

insulating material, and a ventilated cavity on the cold side. Conditions favourable to condensation are particularly important in flat roofs, and this calls for extra care in (a) limiting entering moisture by taking great care with the vapour check and (b) ensuring that through ventilation can take place.

The recipe for dealing with (a) is to use polythene sheet not less than 250 gauge, with all joints taped or otherwise sealed; or to use plasterboard which incorporates a vapour check, taking great care where boards join.

The minimum channel for (b) is 50 mm (2 in). Thus, if the joists are 150 mm (6 in) deep, the maximum depth of insulation is to be 100 mm (4 in) kept at the *bottom* of the channel. The ends of each channel must be vented, which might call for air bricks at each end, protected against entry of birds and large insects. Rather than a continuous line of air bricks, joists may be drilled (not notched) near their ends to allow air to travel between channels. Or easier, if the construction allows, holes or slots may be made in the soffit, to a minimum equal to a 25 mm continuous slot along each side.

Roofs which have one end of the joists built into another structure cannot have channels open at each end. In such cases steps must be taken to ensure air interchange at the closed end, and then to fit roof cowls through which the circulated air may escape.

Warning: channels which are open at one end only will *not* ventilate and are therefore at risk.

There is no need for further detailed analysis of roof types, not because these are in a minority but because an understanding of the principles already outlined will point the way to correct practice. To summarise the main principle, wherever there is moisture that can get at woodwork, there must be a current of air to disperse it. The technique revolves around how that air current is to be ensured, and also to prevent or limit the extent to which moisture can get where it is not wanted.

In that way we can deal successfully with roofs which are clad in copper or zinc, or asbestos cement; those of awkward contours, like Mansards; and so on.

3

Insulating walls and floors

Wall insulation

There is no such thing as an amateur approach to cavity-wall insulation on an existing house. But a little advice to those about to have the job done professionally would not come amiss. For a start, ignore the propaganda, often vicious, which seeks to destroy confidence in the UF (urea-formaldehyde) foam method. If this is done by a competent contractor, preferably a member of the National Cavity Insulation Association, it will conform to the British Standard, and to those rigid requirements imposed by the Government. For far too long it was the Government which would not accept the method. If it is now satisfied, it is not for anyone else to cast doubts.

Accepting the total validity of all three methods, UF foam, blown mineral fibre and blown polystyrene beads, there is little to choose between them for thermal efficiency. Choosing, from reputable people, the best price for the job would be satisfactory. In certain circumstances, mineral fibre might be preferred for its continuing mobility (it can be removed or added to during alterations) and for the fact that it is independent of any on-site physical changes such as are inherent in foam. Meanwhile, however, foam leads the market.

It should be added, however, that if you are building or having built a part or the whole of a house, you *can* do your own cavity insulation, using mineral-fibre slabs as building proceeds. Eurisol UK (see later) will let you have instructions. See *Figure 3.1.*

With solid-wall construction, another method of insulation must be found. It means insulating either the outside or the inside, or perhaps both, so let us deal with the outside first.

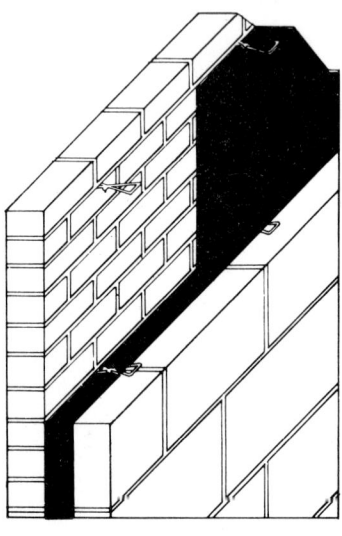

Figure 3.1. Applying rigid insulation as the wall is being built (courtesy Eurisol UK)

The normal process of heat loss through a building structure, that which would occur under laboratory conditions, is greatly assisted by wind and rain, both of which whisk heat away faster. Consequently an elementary form of treatment is that which seeks to exclude wind and rain from an outside wall. Among treatments possible to the handyman is covering the wall by tile or slate hanging, or by barge boarding. Extending the roof overhang also helps. In very windy areas some trees or a strong high fence to deflect winds upwards can be effective.

In a more scientific way, it is now possible to have the entire wall sheathed in a new synthetic skin to a depth of about 100 mm (4 in). This also brings its own decorative colour scheme. It is quite expensive since it must be professionally done and involves, generally, extending the 'throw'

22

of the roof, with guttering and piping, and adjustment around windows. For details ask the External Wall Insulation Association (for address, see Chapter 14).

Internal versus external wall insulation
The difference between outdoor and indoor insulation goes deeper than mere position. Let us take two extreme cases to illustrate the point.

If we own a cottage which we visit for a few hours most weeks, we would be best served by insulating the insides of walls. If, however, we are housebound and therefore in continual residence, there would be a stronger argument for external wall insulation.

The reason for all this is as follows. In the case of the cottage, we need to be able to warm up rooms reasonably quickly – say, in half an hour. Consequently we do not want our warmth being soaked up in the walls of the room, and so we insulate them on the inside.

In the case of continuous occupation, there might actually be some advantage in allowing the walls to absorb heat, always provided that all the heat does not go right through the walls to the outside. For if there is a reservoir of warmth in the walls, the house will stay warm for quite a time even if the heating appliance is not at work for any reason, for example overnight. The inner-lining method is, of course, also useful when there is continuous occupation. The reader will see that cavity insulation is not only half-way between these two but offers no choice.

Internal lining
The only method open to the handyman is inner lining, and we will now examine this. Inner linings take up space. However, this is rarely more than 100 mm (4 in) per wall, in a room which is measured in metres, and a room is rarely so small as to justify objections to anything so beneficial. It has been suggested that a method should be used similar to that for a pitched roof, with an air gap behind the insulation. But that really does use up space, without real justification.

The first job should be to examine the outside wall and to make sure that it is in good condition, bricks or stone and joints. Repoint and patch as necessary. It is then a good idea to apply a coat (or as directed) of a waterproofing solution such as Feb or Lillington. This should be done in dry weather and after a dry spell, to avoid trapping water inside the wall. If you ever had thoughts of doing more, such as cladding the wall in rough cast, or colouring it with Snowcem or similar, this is a good time, for it all helps the waterproofing.

The next job is to waterproof the indoor surface, so that if any moisture should penetrate the wall, the only way it can go is back out again, not into the insulation. A rough and ready way to waterproof is to paint the wall with gloss paint – a good way to use those nearly empty tins. Better is to use a bitumastic paint, or a compounded substance such as Synthaprufe. Instead, or in addition, you may 'paper' the wall with a waterproof substance, and the adhesive, even though it may be only heavy-duty wallpaper paste, must be continuously spread, not put on in dabs. A suitable substance to use as paper is polythene film of about 250 gauge. In all cases the 'paper' must be continuous, with overlapping joins and carried into sides, top and bottom edges.

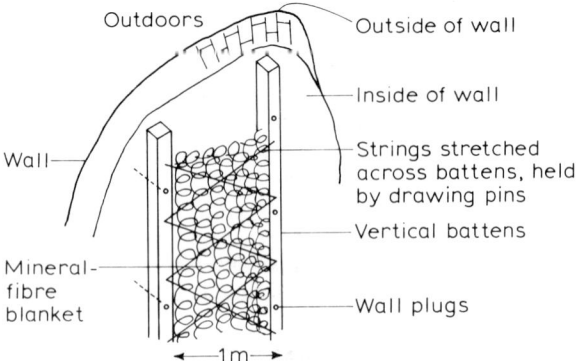

Figure 3.2. Inner lining a wall

Next decide what thickness of insulation you can afford, in cost and in space. Let us think of a minimum of 50 mm (2 in) (see Chapter 1). This will be carried in channels of the chosen depth, formed by fixing vertical battens to the wall with wall

24

plugs and wood screws. The battens will be, for stability, 50 × 36 mm (2 × 1½ in) fixed on edge, and at roughly metre intervals. *Note:* if using a flexible cladding material such as hardboard, a closer pitch is recommended.

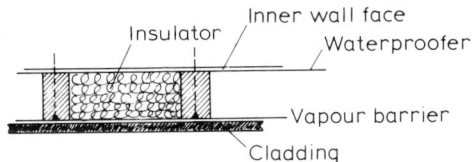

Figure 3.3. Plan view of treated wall section

The insulator may be mineral-fibre blanket such as that sold for loft insulation. It should be cut to be a snug fit in the 50 mm deep channels, in which it may be held during construction by coarse dabs of adhesive on the wall, or by strings stretched across the front edges of battens, with drawing pins. When joining lengths in a channel, push them well together to avoid leaving any gap.

Alternatively, the insulator may be a 25 mm foamed poly-styrene slab, a rigid material which can be cut to size and, as before, made a tight fit in the channels. Butt the edges together. And, since each channel will take two thicknesses, put in the second so as to stagger the joints.

After the insulator must come the vapour barrier or check, to prevent room moisture from getting at the insulation and rendering it ineffective. The cheapest way to do this is to cover the whole job with a polythene sheet, again making sure that joins in the sheet are rendered vapour-proof.

Now we are ready for the final cladding, and the choice is wide. There is hardboard; and plasterboard, which will add to the insulating value; and proprietary versions of plaster-board which are even more insulating; or plywood, which can also be used as a decorative finish, particularly if using a 'faced' type, e.g. beech- or oak-faced; or you may make a real panelling job, using lightweight tongued and grooved boards with woodworker's adhesive in the grooves.

The sandwich method just described is of considerable value in deadening the worst effects of noisy neighbours,

25

when applied to a party wall. It still uses mineral fibre, but if using foamed plastics, choose polyurethane, not polystyrene. An extra benefit may be obtained by using the cladding as the back wall of a built-in cupboard or, in bedrooms, wardrobe, and if kept well-filled it works well.

Last in order of preference for wall lining is expanded polystyrene tiles or sheet. Using a special adhesive sold for the job, these can be stuck straight on to walls, closely fitting together, and for a better job stick a second course on top of the first. It is preferable to buy fire-retardant-treated tiles. This substance has little mechanical strength and dents on impact, but with normal use and care comes to no harm. As a method of insulation this is relatively inexpensive and moderately effective and is certainly better than nothing. It is better, too, than those proprietary wallpapers which incorporate a very thin layer of much the same substance, for the elementary reason that effectiveness is directly related to thickness. That is why two layers are better than one, and three better than two.

Preventing condensation

Condensation is a modern disease. It was rare in the days when open fireplaces evacuated a vast excess of air from rooms, causing ample ventilation. In place of that we now have restricted flue throats, flueless rooms with radiators, and a conscious concern for draught-stopping, all adding up to almost no ventilation. That is why moist air accumulates, and finding a cold wall or window, deposits moisture as condensation, leading to mould growths, flapping wallpaper and so on.

There is a cure for condensation, usually requiring three measures to be taken. The first is the insulation we have been discussing, since it will tend to retard the cooling of outside walls. The second is ventilation, which must be restored but in a controlled manner. All heat-loss calculations make allowance for at least one complete change of room air per hour and it is up to us to see that it can occur. This might mean bringing back the air brick into fashion, though there is more to be said for an aluminium ventilator plate with adjustable damper (see *Figure 3.4*). It can be a case of opening a window, and some windows now include a top section for ventilation only, a narrow louvred slot which can be closed if required. Or it can be a rotating disc damper which rotates under the influence of air currents.

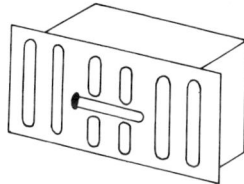

Figure 3.4. Metal-bodied adjustable damper to be fitted into a wall

It must be appreciated, however, that all simple devices of this kind will spend as much time admitting as discharging air, if the wind blows on them. One use of the damper is to modify gale-force winds. A moderate influx of air is just as effective in purging pockets of moist air as an efflux is, since it must be assumed that the entering air finds its way out by some route and ventilation occurs.

But there is a way to put the whole business on a scientific basis, and that is to fit an electric extraction fan, and control it by an humidistat. By this, when the moisture level in air reaches a certain level, the fan starts and runs until that air is purged and the moisture level has fallen. It is useful for anywhere but bedrooms, where the faint click would be disturbing.

A general caution about all forms of ventilator. If not placed in the right position they are wasted. We referred to total room air change, and this envisages a sweeping effect which starts where air enters, and ends at the opposite side or end of the room. It must be obvious that a ventilator placed near a door or window could take air from the open or leaking door or window and discharge it, without ever affecting the main part of the room.

The third rule to avoid condensation is to have less moisture. We supply most of it, so we ought to be able to control it. There are certain activities, like breathing, which

are best not interfered with! But the kitchen and the bathroom have a large turnover in not only water but also water
vapour as steam. The objective is not to alter the amounts,
but to redirect them to outdoors. So keep the door into the

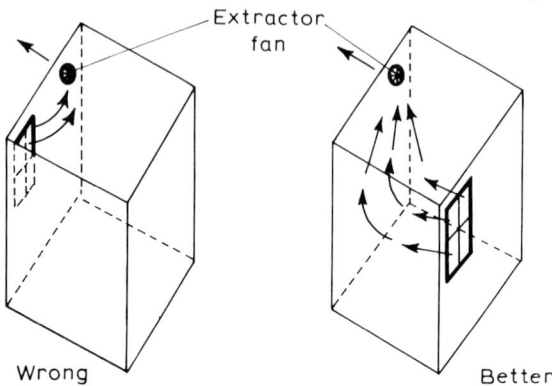

Figure 3.5. Relationship between extractor fan and an openable
window

house shut when working in the kitchen, and let it ventilate
direct to outdoors. This is the room well-suited to an electrical fan, which can either be run at intervals or, if it is a
relatively low-powered model, run continuously during
working times. Position is again important. Fans are often put
near windows, simply because an outer wall has limited area.
A fan adjacent to a window would be wasted if the window
were open as well.

Now a caution about fans in kitchens. A fan administers a
powered sucking effect on the room, leading to a slight
vacuum. This attempts to fill by drawing on every available
source. If your kitchen contains an appliance such as a gas,
oil or solid-fuel boiler with a conventional flue or chimney,
the vacuum will draw flue gases *down* the chimney. In the
case of a gas-fired boiler, this will not become apparent until
headaches, dizziness and possibly worse affect the occupant.
So if you have a fan plus a flued appliance in the room, you
must increase considerably the means of fresh air inlet into

28

the room. This usually means increasing the provision already made. You need to know by how much, and in the case of fuels which emit visible fumes you can see when it becomes safe. But with gas it is best to ask the gas people to run a spillage test for you, which they will be glad to do in the interests of safety. They will then advise you about adequate air inlet size.

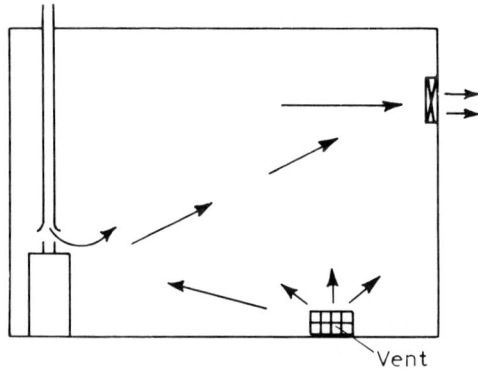

Figure 3.6. The result of extractor fan pulling more air than the ventilator can supply

Although we are the architects of our own misfortunes in the matter of condensation, this is through ignorance, and is sometimes almost unavoidable. It is always worth remembering the large numbers of poor people, many of them pensioners, who in the 1960s were moved into new high-rise flats with electric underfloor heating. The idea was well-meaning and hygienic, but unrealistic. The tenants soon found that they could not afford the electric heating, so they went out and bought paraffin stoves. Not long after that they began complaining about the poor workmanship of the new flats, walls streaming with water, moulds growing, clothes rotting in wardrobes. The press pounced upon those details, and a bad time was had by all. But very few people pointed out, and nobody believed them, least of all the tenants themselves, that almost all the troubles began at the paraffin stoves. They

29

are flueless, and on the basis that each gallon of paraffin produces roughly a gallon of water as vapour, that is a lot of water going into the air, in a dwelling not well insulated, not at all well warmed, and probably hardly ventilated at all.

Local papers show that local councils are still being attacked for the so-called shocking conditions in property rented to lower-paid people, and still nobody can get the message over. Let us make it quite clear to our readers that they should *never* have flueless appliances, for gas or for paraffin. If you want a portable appliance, use an electric one, even though the running cost is higher. There are no side effects.

On the general subject of flues and ventilation, consider the possibilities inherent in a disused chimney. There are often good reasons for capping it off. But if it could be securely cowled instead, either the bricked-up fireplace will create ventilation via the compulsory slot in the brickwork or panel; or, since a high-level vent is often more effective, an air brick may be let into the flue near ceiling level. Or, for air brick, read adjustable damper. The effectiveness of such an arrangement lies in the evacuative effect caused by the chimney, while its position ensures that it will always work in one direction only, being immune from face-on winds.

Floor insulation

Insulation of the ground floor has always been a low priority, even though it can, in an uninsulated house, account for 15 per cent of the total heat loss. After walls, windows etc. have been treated, the total heat loss will no doubt be greatly reduced, but up to a quarter of what is now lost could be going out through the floor. That is something worth investigating.

The standard British ground floor is still the suspended timber one. An essential condition of its well-being is the circulation of air beneath it – air which is cold, usually moist, splendidly suited to carry away the greatest amount of heat. Anyone who has a troublesome floor, e.g. dry rot or woodworm, would do well to consider changing to concrete with built-in insulation. Similarly anyone having a house built should examine this option. In that case all pipe and cable runs must be pre-planned and ducts left for the services to run in. The general scheme of an insulated concrete floor is shown in *Figure 3.7*. The insulation material is high-density mineral-fibre slab capable of withstanding the forces of compression and impact loading.

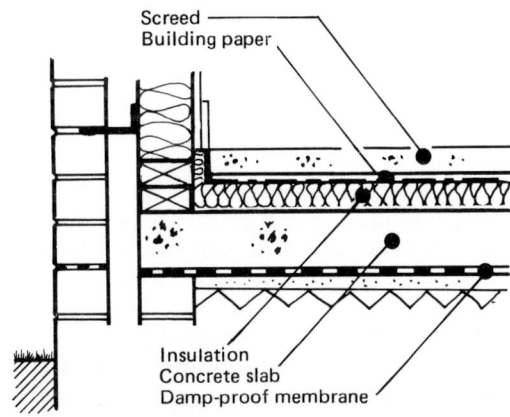

Figure 3.7. Insulation of a concrete floor (courtesy Eurisol UK)

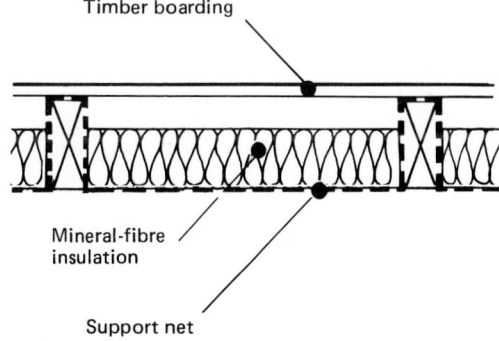

Figure 3.8. Insulation of a suspended timber ground floor (courtesy Eurisol UK)

Meantime, most of us have to keep our timber floors, and there are two ways of dealing with these: easy and hard. And as you might expect, the harder one gives better results. We will deal with this first.

The hard method
Figure 3.8 shows the essentials, close-fitting sections of mineral-fibre mat wedged into the channels between joists, and supported from below by an extremely openwork material, conveniently polypropylene netting tacked or stapled to each joist underside. If you recall what we did with pitched and flat roofs you will see one important difference here – no vapour check or barrier. For reasons too complex to give here, an open-structure insulator, of which mineral fibre is the best example, will allow the free passage of water vapour but not of air. This quality is essential if we are to insulate yet run no risk of dry rot, for it means that the water vapour coming through the floor in the normal manner will be passed on, through the insulation, to be taken away by the air circulating beneath the floor.

To fit this insulation, you should proceed as follows:

1. Lift all floor boards. This is what makes it the harder method.
2. Paint all joists and the undersides of floor boards with at least two, or still better three, coats of an anti-dry-rot, anti-woodworm solution and allow to dry.
3. Meanwhile, make sure that all air bricks are clear and operative so that a good underfloor draught is assured.
4. Tack or staple the netting to the undersides of joists. At the final joist it will be necessary to allow the right length and tack it, from indoors, up the side of the joists.
5. Cut (if necessary) and fit mineral-fibre mat, preferably not less than 100 mm (4 in) thick, to be a good interference fit within each joist channel, and sitting down on the netting. Take special care with the joining of each section of mat, applying some pressure to jam them together to prevent any gaps through which air could pass.

6. Make quite sure that the mat sections are wholly wall-to-wall. There must be no gaps for air to circulate inside the mat/floor board space, for that would nullify the insulation value.

7. Finally, refit all floor boards.

Note: Steps 1 and 7 can be avoided if there is sufficient working space below the floor. In that case access can be had by letting in a trap door instead of removing all the boards.

An easier method

The other way to deal with floors would suit those who cannot or prefer not to lift floor boards, and also those who live in rented accommodation. Let us take for granted that you have made sure of a good under-floor draught, for a start. A wry proof of this might be the strength of the draughts which come up through the floor boards, a serious source of heat loss and the first thing to be dealt with.

1. Examine the bare floor carefully, and make a chalk mark wherever you find a draught coming up. If there are any really bad boards, this is the time to renew them. For the most part, however, filling is sufficient. The cheap and easy way to do this is to make up a quantity of papier maché, once used by every child. Tear some newspaper into short strips, put into a basin and pour on boiling water. Mix to a mashy pulp and allow to cool to lukewarm. Then stir in some prepared cellulose wallpaper paste (or plain flour) and again mix until it is like a light dough.

The mixture is now ready to use. Work it well into every crack, using a flat-bladed instrument (a wood chisel for instance) and leave to dry. When dry, sandpaper each place down to floor level. You can paint or stain it if you wish.

2. Next, cover the entire floor area in hardboard, rough side up. Leave the sheets in the room for at least 24 hours before fastening them down. Then, make a careful job of each join, and tape over the joins when done. Tape also where the hardboard meets the skirting boards.

3. A good-quality underfelt comes next, to be laid evenly over the entire area which will be covered by carpet. It is not usual to fasten this down.

4. This is followed by the carpet. For insulation purposes this should be the best quality you can afford. The sheep in winter keeps warm with a long and tangled fleece, not a short bristly one!

5. It might sound extravagant to suggest rugs as well. But for thermal insulation every bit counts, and many prudent people would welcome rugs to take some of the wear off the expensive carpet.

It will doubtless be obvious that the method described is not only thermally effective but also makes for a very comfortable way of living.

Perhaps we should now add that each step, from hardboard on, does its bit but if circumstances dictate, you can vary the procedure as follows.

1. Hardboard. This can be left out, provided that the draught proofing is done but not otherwise. Or, if the hardboard is needed as the final finish (e.g. as a border around a square carpet), then it should be laid shiny side up. More than that, you should buy the Royal or similar variety to take surface wear.

2. In a kitchen, for instance, a carpet would be out of place. This is the place for hardboard with cork tiles stuck to it. Or you might even use vinyl tiles. But in each case take particular care to join the hardboard so that spilled water cannot get beneath the hardboard sheets.

The treatment described above will probably give a thicker floor covering than existed before, but not enough to affect such fixed heights as skirting-fitted power points. There is, however, one point to watch for, in any room which houses a fuel-consuming appliance, such as a solid fuel fire or a conventionally-flued gas fire. All such appliances must have an adequate air supply if dangerous conditions are to be avoided. If most of the air used to come in under the door, this might be affected. If there is the slightest doubt, remove the door and saw 25 mm (1 in) off the *top* of the door. Here it will be unaffected by changes in floor level, and will diffuse the air current which previously caused a cold draught at floor level.

Insulating other floors

The foregoing notes deal with ground floors, but we ought to give thought to other floors which exhibit similar problems. First, though, let us point out that insulation is not generally applied under floors at higher than ground level, when the room beneath is inhabited. That would deny the occupant some free warmth from below. *But* if the people below are unacceptably noisy, one method of damping down their noise is to pack the interfloor space with the same mineral wool which is used for heat insulation, with perhaps a layer or two of corrugated cardboard in it to diversify. Also use the maximum of heavy soft floor covering: carpet, rugs etc.

There are cases where there is no inhabited space beneath. Take a flat over a lock-up shop or warehouse. It should be treated as for sound insulation, i.e. as much insulation as is practicable, but no vapour barrier, in the interfloor space.

An even more common example is the bedroom built out over a car port, or over a front porch. The area beneath the floor is protected from rain but from very little else. The method of insulation is similar to that described for ground floors, with insulation in the lower part of the joist channels supported by polypropylene net. One difference is that in deference to wind and other action the insulation should be tight-fitting resin-bonded panels of glass fibre, and the supporting net had better be close-mesh wire netting. The second role of the resin-bonded material is to keep insects out.

4

Insulating windows and doors

Windows

Estimates of heat loss from windows, as a proportion of the total heat loss, vary from about 8 to 20 per cent, and of course the estimates are right. Some houses have more windows than others. But in any case this is not a priority item, and we have to look for reasons why window insulation attracts so much attention and costs so much.

On reason is that it draws attention to itself. Because of the high heat loss per unit of area, a single-glazed window is a natural target for condensation. And condensation is a ready source of dampness, moulds and bad smells, loosened wallpaper, all unpleasant things which people feel should have gone for good when they put in central heating.

Another reason may be that (often because of long-term condensation) wooden window frames have shrunk or rotted, need renewing and add to the heat loss with air leakage. We considered the mechanics of condensation in the last chapter, so we know how to lessen its impact. In thermal terms, double glazing properly done will halve the rate of heat loss over the window area. The saving of up to 10 per cent of the total is worthwhile, so we can begin by deciding what double glazing is, and is not.

For a start, the glass is there only to see through. The insulator is the trapped slice of air between the two panes, and at maximum efficiency it is still, and dry, and very close to 18 mm (¾ in) thick. If it is other than this, the efficiency falls off and less heat is saved. It happens that the same kind of arrangement, but with a 100 mm (4 in) gap, is very good as a sound insulator, for instance for a window overlooking a busy street. But you cannot have it both ways. The better for sound the worse for heat, and vice versa. Those sellers who claim a high performance in both categories for their product are being less than realistic. Fortunately we have got rid of most of the rogues who claim that by fitting double glazing you will halve your total heat losses. The best defence is enough knowledge to be critical

Proprietary double glazing
If you buy proprietary double-glazed units you can take for granted the still and the dry. But you will often find that the gap is less or more than 18 mm, often for production reasons. But reasons are not your concern, whereas efficiency is.

The frames of proprietary units are important too. For a long time aluminium was used exclusively, and this has many advantages. It does not corrode and is virtually everlasting. It does not need painting, but will take paint if colour change is needed. Or colour can be built in by anodising. However, there was a growing tide of discontent at the amount of condensation coming, not now from windows, but from frames. The metal is a very good conductor, and the bogey of condensation did not go away. Aluminium frames are still widely made, with the old advantages, and quite often something can be done to ease the problem. For instance, an insulator such as a rubber strip can be stuck to the frames, using adhesive over the whole area. Or frames can be drilled and tapped and lined with timber strip bolted on.

A section of the trade turned over to plastics, and there is now a thriving market in this material. The same advantages are claimed, such as fast colour without painting, and latterly an assurance of no surface deterioration. This is probably true, and memories of crazing of plastics by the action of ultra-violet light no longer relevant. To make sure of the point we would expect a manufacturer to give an unconditional very long-term guarantee against cracking, crazing or deterioration of the material.

The most promising development has arrived by way of a compromise. This allows us to have the advantages of aluminium frames with their freedom from corrosion and absence of maintenance, and, by means of a plastics insulating infill, much lower conductivity and therefore a reduction in condensation. *Figure 4.1* shows one of the foremost of this new and increasingly popular type.

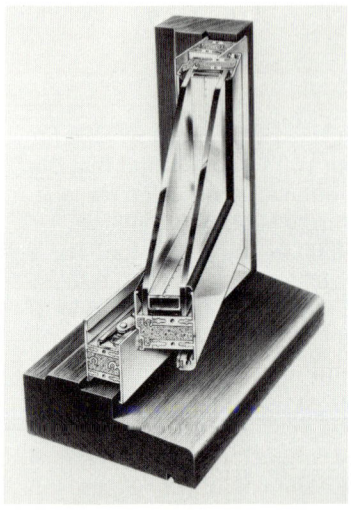

Figure 4.1. Section through Thermastor double-glazed opening window showing the insulating material within the frame, positive gap fixing, and sealing of panes and of closed window

Commercially, double glazing is divided into that which is sold as a commodity, and that which is sold only as an installed package. For those who choose the second, these notes will at least offer a means of identifying the product with the best specification.

Fitting a proprietary double-glazed unit is a task within the capabilities of most people who can do a moderate job of carpentry. This is partly because, for reasons mentioned, main frames often need repair, and should be in very good condition before a new window is fitted. The other reason may be that a small amount of construction is called for, usually letting an extra piece of wood into the frame to reduce a dimension.

We cannot give details, since each brand of window has its own rules. But these are made quite clear, either by the seller or by a representative who will call to verify the measurements; and probably in writing with illustrations when the windows arrive. But there are certain general rules which apply.

1. Make sure that the main window frame is in good condition. If not, see to it before attempting to fit a new window. Remember that a frame member which is recessed into a brick wall may have deteriorated within the cavity. If in doubt, either fit or have fitted a new frame.

2. If you do your own measuring, make sure that you understand the instructions. Since most replacement windows have stepwise dimensions, on no account buy one larger than the opening. You must always buy a size down and fill out the frame to suit it.

3. Check whether the main fitting operation, e.g. screwing a metal frame to your wooden frame, has to be done from indoors or out. You might need to make special arrangements for an outdoor job above ground-floor level.

4. Be most particular about the final assembly, of window to frame. If a jointing compound is supplied, use it. If not, lay a thin even smear of a mastic jointing along the area of contact on the frame. Above all, tighten down screws gradually and in cross rotation, much as you do when changing a car wheel. Aluminium windows frames are easily warped and then they never fit.

DIY double glazing
Now we come to the entirely home-made double glazing. It is quite difficult with metal frames because unlike wood they will not take screws but must be drilled and tapped or bolted through. These are specialised jobs, requiring the right equipment, but apart from that the methods are the same.

The difference between the two methods is in the way the second pane is secured to the window. We may use a proprietary sealing strip, which is a plastic channel screwed to the frame with the glass firmly held in the channel. It has

39

the benefit of being removable, but great care must be taken to mitre the corners accurately, and some jointing compound is advised there. The other method is more self-reliant. The new pane is bedded on a jointing medium and held in position by a wooden fillet as shown in *Figure 4.2*. The jointing medium may vary from double-sided adhesive tape for very good surfaces, to a thin line of putty.

In both cases begin by ensuring that the frame is sound. Use a pane of glass large enough to rest comfortably, with its jointing, on the frame, and leaving ample room for the

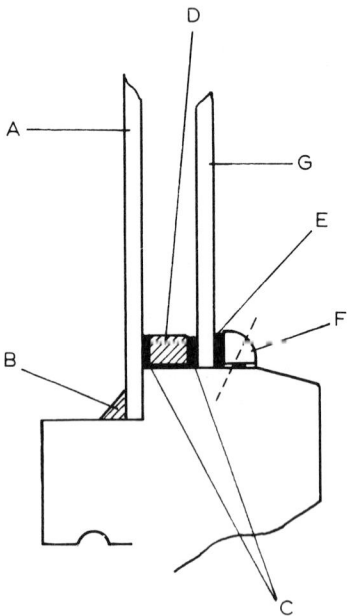

Figure 4.2. Adding an extra pane to an existing window. Existing pane (A) held in place and sealed by putty (B). (C) strip jointing of thin rubber, mastic or similar. (D) spacer, conveniently wood fillet 18 × 12 mm. (E) mastic or putty seal, protected by wood bead (F). (G) new pane

edging strip of whatever type. When putting the pane on the window choose a very warm and dry day, or if it is winter, beam an electric fire on to the job from fairly near. The object is to trap dry air, otherwise internal condensation will be seen. It is not always practicable to manage the 18 mm gap called for but the consolation is that some insulation is better than none. Any method of inserting a pane into the frame to get the right gap turns out to be very hard to make airtight.

Clear plastics can often be used instead of glass, though they sometimes do not pass so much light, a consideration on winter days. A very useful application is in bathrooms and other places where the glass is not see-through. For these any of the figured plastics sheets may be used without any disadvantage. The rules for size and jointing are the same, the method of securing is easier. Plastics can themselves be drilled and wood screws used, with or without edging strips, to hold the sheets in place.

The second pane is no longer the only form of window insulation. We now have reflective insulation. This is a special skin which can be stuck to a window and it acts by reflecting warmth back into the room instead of letting it escape. We can do no more than mention this and advise you to look out for it. At present it has to be professionally installed, and is mainly applied to commercial and industrial buildings. But these good ideas have a way of acquiring a domestic face in due course.

Doors

There are three kinds of treatment for doors.

1. Glazed doors must have their glazed area treated as a window, as above.
2. The door itself, particularly if the panelled type, allows a lot of heat through. Any door can be lined on the inside, over the timber area, with hardboard.

a) First inspect the door carefully for soundness, inside and out. Pay special attention to joints, e.g. around the edges of panels, and work plastic wood or similar filler into

41

suspect joins. Paint and varnish the outside, and treat the inside with a Cuprinol or similar wood preservative.

b) Fill panels up to flush with mineral-wool blanket, held in place with a dab of adhesive.

c) On other, flush, surfaces where there is no room for a packing-type insulator, use a reflective type, aluminium kitchen foil being convenient. Again dabs of stiff adhesive are helpful.

d) Measure hardboard to cover the timber area except where the door fits into the jamb on closing. See *Figure 4.3.*

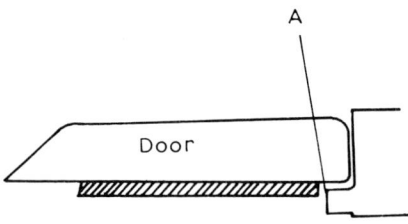

Figure 4.3. Lining a door on the inside. Note that the cladding stops at A

e) Secure the hardboard to the door, using ample brads over the main door structural members, with stouter fasteners along the edges. A timber fillet will act as a holder as well as improving the appearance.

3. The third trouble with doors is usually that they, or their frames, leak. They must be made airtight when shut. If this is a case for joinery, we have already seen to it. But perhaps it needs draughtstripping, and there is a wonderful variety on the market. There are two things to note about it.

a) It must be continuous around the perimeter of the door.

b) It must be the right thickness. A common mistake is to buy, for a door in ordinary good condition, a too-thick seal, on which the door will not even close.

Draught excluders are made in a variety of materials, at prices which roughly indicate their value.

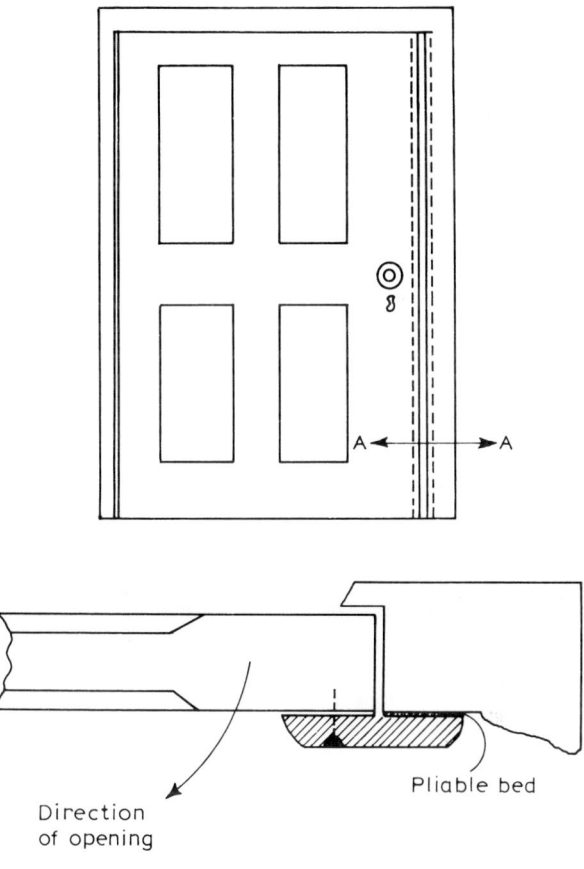

Direction
of opening

Pliable bed

Enlarged section on AA

Figure 4.4. A strip of wood or metal added to the door can
reduce draughts through the gap

There are foamed plastics, not to be highly recommended
since they have a short active life. The effects of wear on
moving surfaces, and of ultra-violet light, probably mean
renewal annually. In the category of soft pliable items soft
rubber is much better.

Nylon pile stuck to an incompressible backing is good but has a minimum thickness which might be too much for narrow gaps. Check before buying.

Bronze strip, in a flexible V-section which acts like a spring in closing a gap, is hard-wearing and effective. When working properly it calls for some physical effort to close a door, and where infirm people are involved, they should be allowed to try a sample fixing.

A method of dealing with doors which is not invisible, and therefore not to everyone's liking, can nevertheless be very effective and even made into a visually acceptable feature. As *Figure 4.4* shows, it is a strip of wood or metal added to the door, which in effect closes a lid on the gap through which draughts come. Doors do not always close flush with their frames, as *Figure 4.4* section shows, but this simply calls for either rebating of the strip or packing out with a full-length fillet, depending upon which member is proud. The pliable material which is the sealer may be rubber, synthetic rubber or backed nylon, and some account should be taken of the material to be used, and its thickness, when setting the strip.

A final note on doors. The best thing that can happen to any outside door is to have a closed porch built around it. That is in addition to the treatment just outlined.

Wind and rain both take away heat from warm bodies, and in addition cold wind blows in through an open door and upsets the thermal balance for a long time. A closed porch saves all that trouble, and gives a handy place to store umbrellas.

Failing that, fit a storm door, which is a second door just outside the original one.

The 'greenhouse' effect

This is very important and worth understanding, because so much depends upon it.

Radiant warmth is a form of wave motion, just like light. The wavelength varies according to the temperature of the radiant source, being shorter for higher temperatures. Thus, radiation from the sun has a very short wavelength. Ordinary glass can be said to act as a filter of warmth. It will let very short wavelengths through, and so we feel the heat of the sun through a window.

The sun's warmth enters the room, and warms up furniture, walls, and so on, but only to a very low temperature (compared with that of the sun, anyway). The warmed objects then emit their own radiation, but of course this is long-wave radiation. And glass, the filter, will not let it through. So that warmth stays in the room. And in consequence the room, or greenhouse, becomes hotter and hotter in continuous sunshine.

The net result, called solar gain, is beneficial to some, but is a real problem to others, for instance those who work in office blocks with glass walls. It is helpful to solar heating collectors, and in general to greenhouses.

5

Draughts and ventilation

The general subject of draughts and ventilation has been touched upon in other chapters, thus proving that it is many-sided and important. There appears to be a built-in paradox in that we try, by central heating, to achieve efficient warming of air, and then we deliberately encourage some of it to escape. In fact regular dilution is necessary, since air trapped in living quarters becomes increasingly impure and laden with moisture.

In the days of open coal fires, ventilation was never a problem, for an open chimney can move on average ¼ ton of air every hour, and this has to be replaced from outside the room. In such circumstances draughts were inevitable, and it is not least a reaction against those times which explains the emphasis upon draught-proofing now. Where many a mistake is made is in being too thorough, and virtually blocking-off all points of air entry. This leads to smoking chimneys and, worse, the same phenomenon with gas fires where the 'smoke' is invisible. The mechanism of smoking chimneys is described in the next chapter. The aim should be to get rid of all accidental draughts since these are not under control. Then introduce a source of draught which is under control.

To recapitulate briefly, the principal sources and treatments of draught are:

• the floor, if suspended ground floor. Renew bad boards, fill cracks with papier maché, cover the floor completely with hardboard followed by felt and carpet or similar. Check around skirting boards.

● windows and doors should be given wood surgery if in a bad state. Otherwise, use draughtstrip in one of its many forms.

The main methods of introducing ventilating air are:

● by relieving the *top* of an internal door, usually by cutting off 25 mm (1 in).

● by a ventilating window.

● by a hole cut in a suspended ground floor, and fitted with an adjustable damper. This is usually situated adjacent to the appliance (*Figure 5.1*).

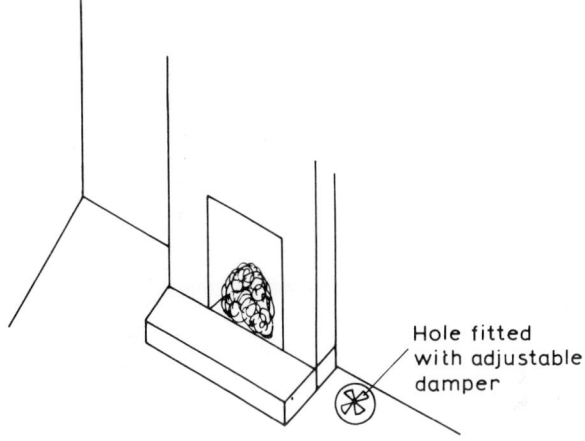

Hole fitted with adjustable damper

Figure 5.1. Combustion air from beneath a suspended floor – ground floor only. This assists underfloor draught as well

● by a hole cut in an outside wall, the air usually conducted in a duct, and an adjustable damper fitted (*Figure 5.2*).

● by a hole cut through a wall into an adjacent room, requiring treatment of the second room.

Remember that holes which lead to the outside allow for about twice as much air change as those which lead to the next room. But they have substantially the same effect irrespective of what fuel is used, though it would seem desirable to apply a rather higher standard to solid-fuel appliances.

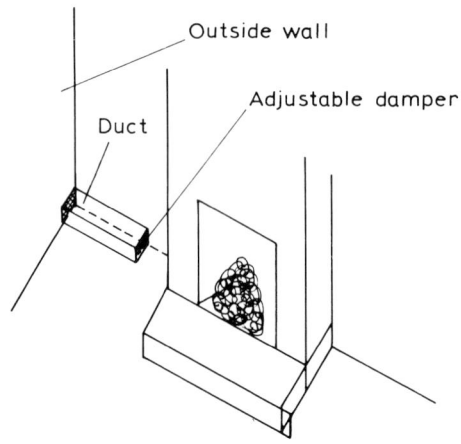

Figure 5.2. Combustion air from outdoors, through the wall and ducted to the fireplace area

A useful guide to the size of the holes is shown below, with minimum values given. High-level vents apply to boiler-type appliances, those which can give off considerable warmth in confined spaces, e.g. under the stairs, unless the air is purged. In the case of balanced-flue appliances, both high- and low-level vents are for air change only, since there is no flue demand for air.

For appliances with an open flue, if air is drawn from the next room:

at high-level permanent free open area $9\,cm^2/kW$ ($1\,in^2/2000$ Btu/h)

at low level $18\,cm^2/kW$ ($1\,in^2/1000$ Btu/h)

If air is drawn from outdoors, halve the above areas.

Note also that if air is taken from another room, that room also must have free access to air.

For appliances with a balanced flue, if air is drawn from the next room:

at high and low level, $9\,cm^2/kW$ ($1\,in^2/2000$ Btu/h)

Where the air is from outdoors, halve the areas shown.

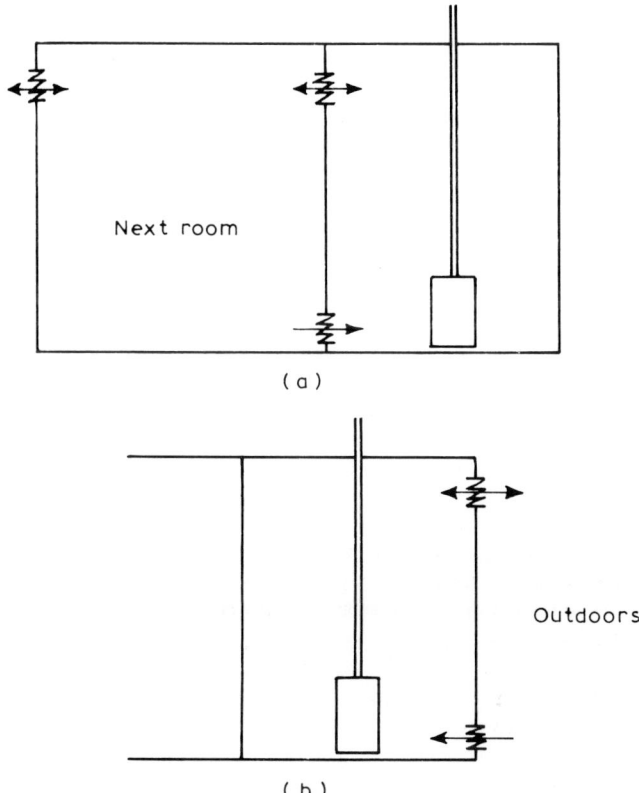

Next room

(a)

Outdoors

(b)

Figure 5.3. Disposition of air entry/discharge points in a room containing an open-flued appliance vented to (a) indoors and (b) outdoors

Notes

1. The appliance ratings shown are the rated outputs, in kW or Btu/h.

2. In the case of open-flued appliances, an important exception to the rule exists where an electrical extractor fan is fitted, as for instance in a kitchen. A greater area must be

provided, the amount to be decided either following a spillage test (for gas) or when the appliance ceases to 'smoke' (solid fuel and oil). Gas technical staff can carry out spillage testing.

3. The areas of holes quoted above are not the areas of excavations e.g. through a wall. It is usual to fit an air brick or a fabricated louvre/damper. You can measure, or ask the maker, what the percentage free area on these devices is, and in most cases it is not better than 30 per cent. Consequently, if for instance your appliance needs a permanent free area of 10 in^2 the area of the excavation will be not less than $10 \times \dfrac{100}{30}$, or more than 30 in^2.

4. Air vents to outdoors and to the subfloor should be screened to keep out larger insects. This too is a restriction of the real area, and should be allowed for, and the screen cleaned from time to time.

5. Air vents to outdoors must not be placed in the vicinity of a balanced flue terminal, where they could admit flue gases into the house.

6. The placing of vents in all the categories usually allows some choice of position. Every effort should be made to avoid promoting draughts across seating areas, whether at floor or at high level. The desirability of a location adjacent to the appliance has been mentioned, and high-level vents should be near the ceiling, for usefulness as well as for comfort.

7. If more than one flued or flueless appliance is in a room, the air vent allowances should be those for the appliance with the greatest demand judged by the rules above.

Flue liners

Until builders went through a silly phase and built houses without chimneys, these vertical tunnels were taken for granted. But most chimneys are more crude than is desirable when serving appliances scientifically designed for high efficiency. Being of large area they are well suited to old open

50

fires and the vast amount of air evacuated by these. But fit a restricted throat appliance and the flow rate becomes a trickle, the travel so slow that cooling and condensation take place half-way up, and damp patches appear on bedroom walls.

The cure, for low-speed travel, for condensation on brickwork, and very often for marginal cases of downdraught, is to fit a chimney liner. This is a tube, of metal, ceramic material or asbestos cement, which connects direct to the flue offtake of the appliance at its lower end, and to an approved pattern terminal at the top end (*Figure 5.4*).

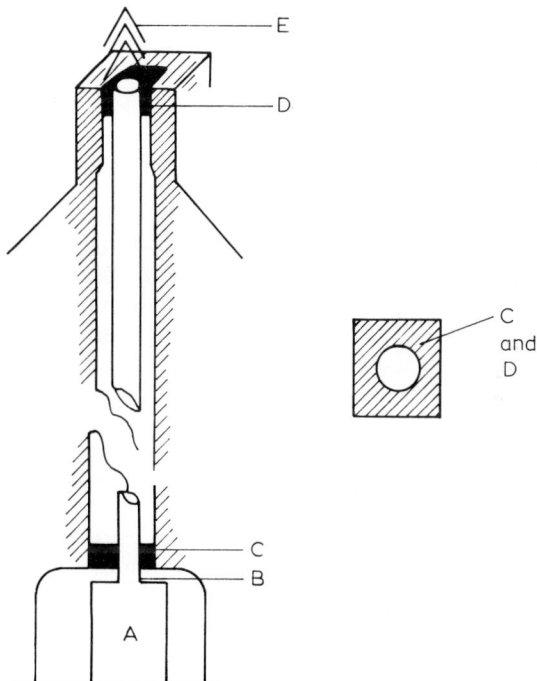

Figure 5.4. The features of a flue liner. (A) is the appliance, sealed into the lower end of the liner at (B). (C) and (D) are means of wholly sealing off the annulus between flue liner and original chimney. (E) is an approved pattern terminal fitted to the liner

The flue liner is, in short, a very good buy; in effect, it is a new chimney on the cheap since it is far cheaper than a new fabricated chimney. Moreover, it does not require Building Dept. sanction, and you do not have to alert the fire brigade and your fire insurer, as you do with newly constructed flues (rightly so, we must add). People do fit their own liners, though all the feed-in is done from the roof ridge. Even if using a flexible liner, the amateur should make sure that his is a straight flue within reason, to avoid difficulties. Pay special attention to jointing at each end. Not only has the liner to be connected, but the annulus – the space between liner and old chimney – must be filled and sealed off at top and bottom, to prevent the movement of air, and possible entry of rain, into that space.

Building a chimney

For situations where there is no available chimney, one may be constructed, and structurally there are two ways. You may use pipe, usually asbestos cement pipe, which unlike metal does not corrode. Or you may use precast blocks and build up a chimney.

In both cases there is a choice, to run the new flue mainly indoors or mainly outdoors. Thermally the first is desirable. It avoids overcooling of the flue gases and deposition of water vapour, and it ensures that such heat loss as does occur is kept indoors and is therefore beneficial. A flue which is run mainly outdoors calls for rigorous insulation, and weatherproofing of the insulation, or in the case of asbestos the use of double-walled pipe.

The advantage of an outdoor pipe lies in its construction. It is relatively free from building restrictions. The indoor pipe is hedged around with conditions, the main one being that it shall be at a distance from structural timbers, which must be framed around where it passes through floors and roof. And it must pass inspection on these matters, for it is a matter of great concern to the fire prevention people and to your own fire insurer. See *Figures 5.5* and *5.6*.

The other form of instant chimney or flue is that which is made of precast blocks and when built up bears a strong

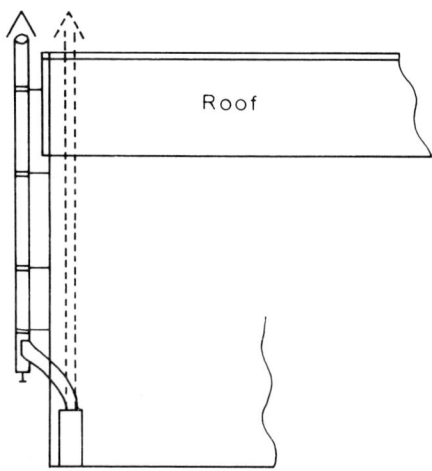

Figure 5.5. An independent flue may be run outdoors (note the condensate trap and drain) or indoors

resemblance to a traditional chimney, whether it is run outside or inside the house. It is less bulky than the old pattern chimney, not least because the flue opening is not any longer 23 × 23 cm (9 × 9 in) but is of a size appropriate to its duty. This type of chimney is now well established and certain makes have earned the approval of the NCB (National Coal Board) or SFAS (Solid Fuel Advisory Service) on behalf of solid fuel, and the Gas Council. Consequently you can expect to get information from an SFAS office or from a gas showroom.

All flues and chimneys, whether traditional or built up, are expected to have an appropriate terminal nowadays, and information about this can be had from the SFAS or gas showroom for approved patterns.

Draught diverters

Gas- and oil-fired appliances which are connected to open flues (not balanced flues, which we shall mention) all have a

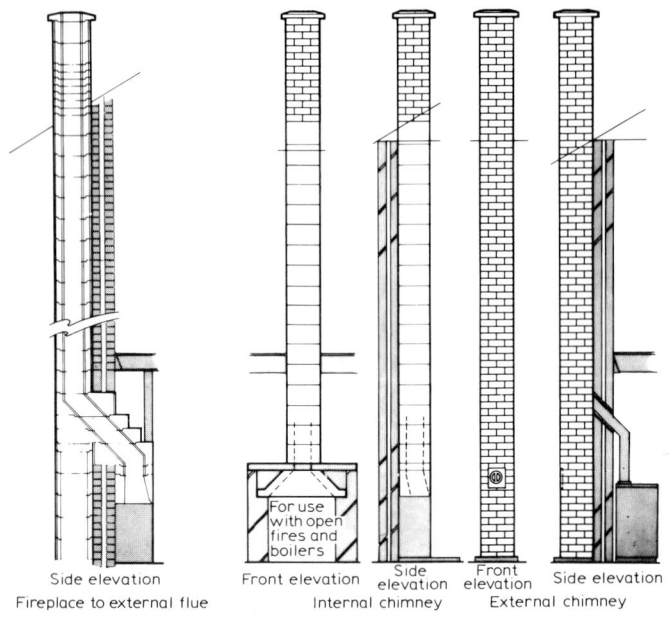

Side elevation
Fireplace to external flue

Front elevation

Side elevation
Internal chimney

Front elevation

Side elevation
External chimney

Figure 5.6. Added-on chimneys made from prefabricated sections, run indoors and outdoors (courtesy True-Crete Ltd.)

draught diverter fitted either near the flue offtake or some-where within the appliance. Only open gas fires are exempt. The device, more properly called a downdraught diverter, does precisely what its name says. A strong downdraught hitting a flame could extinguish it, which at the least would be a nuisance. The diverter ensures that wind blowing the wrong way is diverted away from the flame area, and general-ly out into the room housing the appliance, in which it must be fitted.

At the same time it provides a way of escape for the flue gases which continue to be formed, until the flue, or secon-dary flue, is again working the right way round. A diverter has two other functions. It brings about some dilution of the flue gases. And it acts as a damper on flue pull, which is capable of operating in the opposite way to downdraught with lively

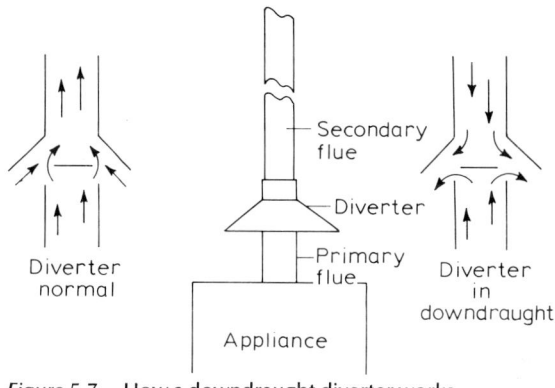

Diverter
normal

Secondary
flue

Diverter

Primary
flue

Appliance

Diverter
in
downdraught

Figure 5.7.   How a downdraught diverter works

favourable winds. An appliance which suffers from down-draught for most of its working time should have its main flue examined.

Balanced flue

We must look now at the balanced flue, which goes with the room-sealed appliance, terms which will be better under-stood by studying *Figure 5.8*. This is really only applicable to

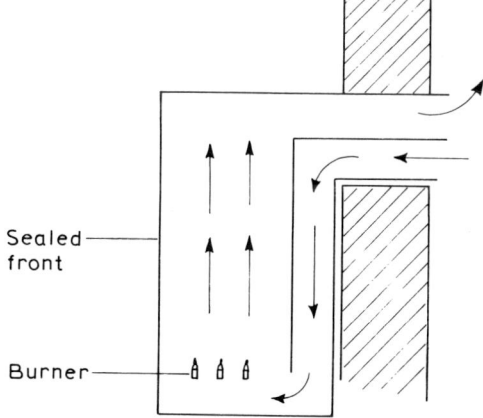

Sealed
front

Burner

Figure 5.8.   Balanced flue in room-sealed appliance

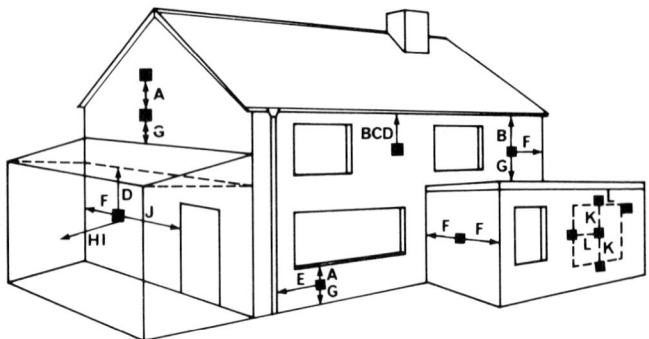

Figure 5.9. Minimum distances of balanced-flue terminal from building features (courtesy British Gas)

	Minimum distance	
Terminal position	*Natural draught*	*Fanned draught*
A Directly below an openable window or other opening, e.g. air brick	300 mm	300 mm
B Below gutters drain or soil pipes	300 mm*	75 mm*
C Below eaves	300 mm*	200 mm*
D Below balconies	600 mm	200 mm
E From vertical drain pipes and soil pipes	75 mm	75 mm
F From internal or external corners	600 mm	300 mm
G Above ground or balcony level	300 mm†	300 mm†
H From a surface facing a terminal	600 mm	600 mm
I From a terminal facing a terminal	600 mm	1200 mm
J From an opening in the car port (e.g. door, window) into dwelling	1200 mm	1200 mm
K Vertically from a terminal on the same wall	1500 mm	1500 mm
L Horizontally from a terminal on the same wall	300 mm	300 mm

* If a terminal is fitted within 850 mm of a plastic gutter or 450 mm of painted eaves an aluminium shield 750 mm long should be fitted to the underside of the gutter immediately beneath the gutter.

† If a terminal is fitted less than 2 m above a balcony, above ground or above a flat roof to which people have access then a suitable terminal guard should be provided.

When considering the termination of a balanced flue in a single-storey extension or car port it must be appreciated that the structure should only consist of a roof and one other wall in addition to the main wall of the dwelling.

From the chart it will be seen that where there is entry into the dwelling from the car port the terminal should be positioned a minimum distance of 1200 mm from the opening, which may be a door or window.

Where there is any deviation from these conditions the balanced-flue installation should be regarded as suspect.

When the distance of the opening is found to be more than 1200 mm then D,F,H,I must also apply where D is the vertical difference below the lowest point of the roof and the top of the terminal.

Siting adjacent balanced-flue terminals

Where the situation demands that it is necessary to site balanced-flue terminals adjacent to one another on the same wall, of which the most severe case is one terminal vertically above the other, it is found in conditions of complete calm a relatively stable plume of combustion products could rise from the lower appliance to vitiate the upper one.

A lesser distance is required horizontally, as in this direction the wind can be expected to disperse combustion products rapidly. The recommended spacing is given in the table and applies to appliances up to 30 kW output.

A sequence of appliances fitted vertically one above the other will give rise to only a negligible cumulative effect which may be ignored.

Se-duct systems

The system consists of a single vertical duct normally of rectangular section, open at the top and base, to which appliances with sealed combustion-chambers are fitted as required, on any floor.

There is a free flow of air through the duct; combustion air for the appliances is drawn from the duct, and the products of combustion are discharged into the duct.

The size of the duct is related to the number and type of appliances connected to it in such a way that the volume of air flowing through the duct is sufficient for satisfactory combustion of each appliance under all conditions of use.

gas appliances, and quite a large proportion of all new appliances are now room-sealed. Appliance efficiency is as high as with open-flue appliances, and there is much more versatility about location. All that a balanced-flue appliance needs is an outside wall, of which there are more than available chimneys. That must be qualified a little, because there are conditions.

A balanced-flue terminal must not be located near an openable window or door, or at low level adjacent to a public right of way. It must be kept away from corners, and nearby walls, bushes etc which could cause wind eddies. The first provision is because flue gases are not meant for breathing, the second because the whole principle depends upon air intake and flue outlet being equally affected by whatever wind is blowing.

The requirements are set out in more detail in a drawing prepared by British Gas and reproduced here by courtesy of British Gas and Benn Bros (*Figure 5.9*). The dwelling incorporates all the limiting factors, and by reference to the tabular chart we can find what the minimum distances must be of balanced-flue terminal from the building feature e.g. window, gutter etc. The rising popularity of the wall heater means that more people will be having more balanced-flue terminals.

The only case not covered by the drawing is that of the terminal which discharges on to a right of way, a potential two-way bother. Some passer-by might complain, or idle people might be tempted to 'post' sticks, stones and worse, in spite of the guard which must be fitted. Generally, if such terminals can be put above head height they will be out of trouble. But local bye-laws could affect the issue and it is best to find out first whether any such bye-law exists.

Balanced-flue appliances come into the category of room-sealed because the entire combustion process is sealed off from the room – always supposing that the sealing has been properly carried out and not subsequently broken. Flue gases leave at the flue terminal, having risen, warmed by the flame. They leave a space into which air flows by the only available route, which is down the air duct from the terminal.

In this way a steady movement is set up which leads to consistent combustion, whatever the weather conditions and force of wind outdoors.

To make a simple mathematical case of it, let us suppose that the flue gases emerge under pressure p, and air enters with suction s. Then the operating force across the appliance is p + s. Now a wind force P blows on the terminal. The flue gases meet resistance, and have force p − P. The air intake is assisted, to force s + P. And the new differential is (p − P) + (s + P) which as you can see equals p + s. So P does not affect the operation.

Room-sealed appliances are always used where Se-ducts and U-ducts act as the flues. But these ducts are found only on specially built, usually tall, buildings, e.g. purpose-built blocks of flats, and ample advice will be given to anyone moving into such a building.

Causes of downdraughts

It was mentioned that a flue liner could assist in marginal cases of downdraught, and that is because it causes greater velocity, hence more powerful ejection, to overcome the forces opposing ejection. But some flues suffer more than marginal downdraught, either permanently or in certain conditions. Let us examine the more common of these.

1. An obstruction such as a hill, high building or tree in the path of the prevailing wind, which causes the wind to lift but to fall again just in time to fall on the flue terminal (see *Figure 5.10*). Rarely can anything be done about the cause, and the only likely cure lies in a special flue terminal, in principle similar to the downdraught diverter. Amid a long catalogue of failed cowl designs the name OH stands out as the one with the highest success record (*Figure 5.11*).

2. It should be self-evident, but we must mention the possibility of obstruction within the chimney itself, whether from a bird's nest or fallen brickwork. A flue brush will discover any blockage, but beware of falling debris.

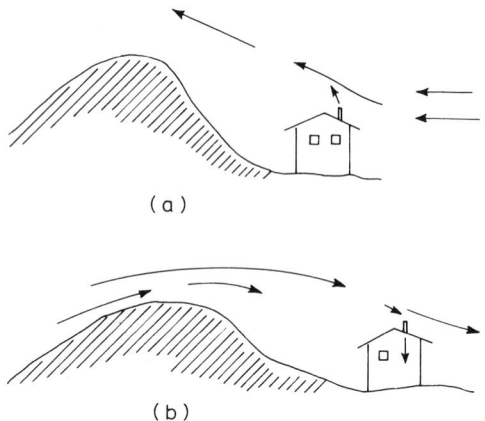

Figure 5.10. How downdraughts can be caused. In (a) the prevailing wind is from the right and there is no problem, but in (b) where it is from the left, the hill causes the wind to rise, then fall on the flue terminal

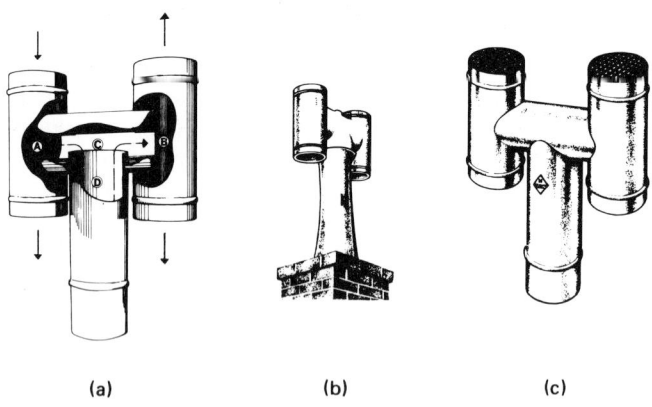

(a)   (b)   (c)

Figure 5.11. The OH Cowl. (a) section showing gas passage through internal ejectors. Right-hand shaft is shown working normally, that on the left responding to a downdraught. (b) is an earthenware general-purpose pot, and (c) is metal, British Gas-approved. These pots and cowls are not recommended for flues larger than 33×22 mm (13×9 in)

3. Since the public became draught-conscious and began applying draught sealer, most downdraughts have been self-inflicted.

Without going into detail, it may be taken that every flue and chimney in its given situation has a natural appetite for gases, whether flue gases or air or both. It will not tolerate more, but will eject the excess and 'smoke'. Nor will it be satisfied with less, being determined to pull from the room just what it needs. What it takes from a room is replaced in the room by air streams from other rooms, windows, doors – all those sources of replenishment which are sealed off during draught stopping.

So the pulling chimney soon begins to create a partial vacuum in the room, and, as we know, nature abhors a vacuum. So the vacuum 'tugs' at the only open source: the chimney. Within the chimney a struggle goes on, the 'pull' pulling one way, the room vacuum the other, resulting in a see-sawing effect. When the room is winning, the chimney smokes. This diminishes the vacuum, allowing the pull to win temporarily, and so on. It explains why a smoking chimney often misbehaves in puffs.

The cause can be proved by leaving a door or window open, and cured by relaxing some of the sealing. But a better remedy is to provide a purpose-made air inlet, which not only allows control over the air admission but enables it to be routed so as to avoid the discomfort of draughts to room occupants. A favourite method is to admit air from beneath the floor or through a side wall, the fire wall, as *Figures 5.1* and *5.2* show. Fit adjustable dampers to these inlets, and also a form of mesh to keep out animal life.

Oversized chimney

Lastly, we come to the chimney which, like many older ones and seen in its extreme case with the ingle, is too big for its modern job. Once a little warmth is applied, a chimney has no preference for flue gas over air but takes what comes easily. As *Figure 5.12* shows, the proportion of opening to fire

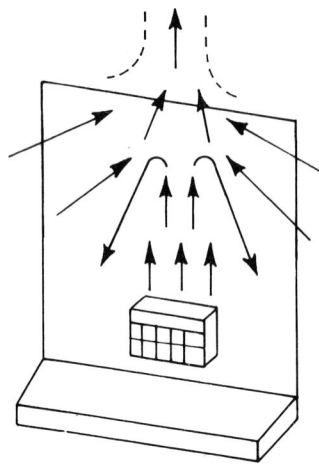

Figure 5.12. Why a small fire smokes in a large opening

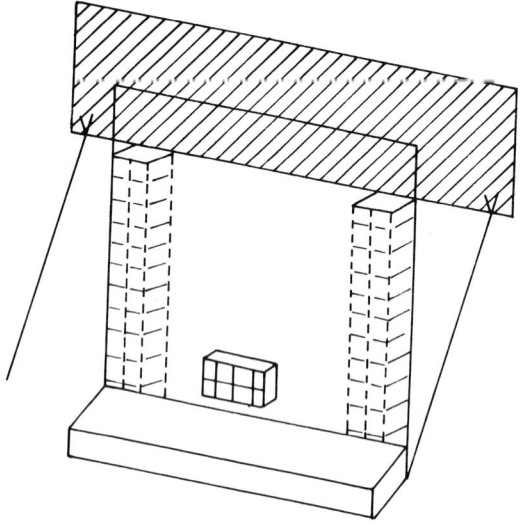

Figure 5.13. Making a trial-size reduction of the opening

makes air more accessible than flue gases, and so the chimney will satisfy its natural appetite, already mentioned, with air. This leaves the flue gases with nowhere to go except out into the room. It smokes.

Knowing the cause we can begin to make a cure, by reducing the size of the opening. This is to an extent empirical and fortunately we can do everything on a temporary basis to start. The width can be reduced by pilling up bricks at each side, with no jointing. And although the final height reduction will be by a canopy, any sheet of metal or asbestos millboard may be supported against the front of the breast, and moved up or down, to find a non-smoking position. Only when the experiments are successful should the new dimensions be made permanent, the height by a canopy, the sides by firebrick jointed with fireclay.

That remedy is good for normal cases of oversizing, as in some houses of the Victorian period. The ingle is really too large to make it a good proposition, and by far the best way to deal with it is to fit a free-standing room heater inside it, with a flue liner running up through the flue (*Figure 5.14*). But

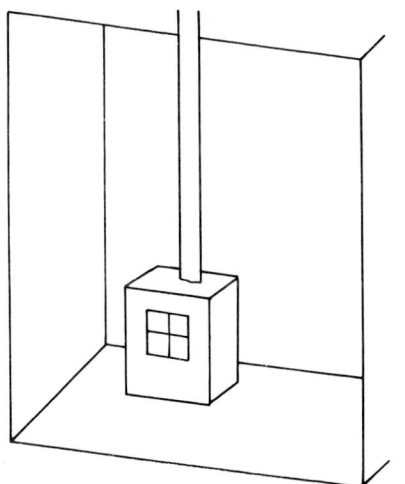

Figure 5.14. Freestanding appliance and flue liner fitted in an ingle

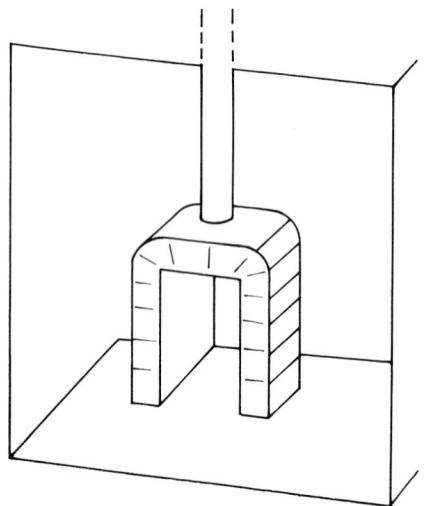

Figure 5.15. Workable size of fireplace
built inside an ingle

we have seen cases of a fireplace built within the confines of
the ingle, with only 9 in side walls, and suitably reduced flue
area, leaving the bulk of the original area still available for
sitting, ornaments etc (*Figure 5.15*).

Kitchens

Kitchens tend to be a special case since they usually house a
gas or other boiler, and very likely have an electric extraction
fan as well. The combination is potentially dangerous. As we
said earlier, a fan is capable of overcoming the flue so as to
ensure that the flue gases come into the kitchen, and
remember that they are invisible. So if you have a boiler/fan
combination, be sure to ask the gas people to do a spillage
test in case you need extra air inlets.

The second practical hint is this. If for any reason you have
to make more area for air to enter the room do it by sawing
25 mm off the *top* of the door. In that way the gap is assured,
even if you buy a thicker carpet.

6

Choice and calculation

Not everyone has a free choice of the method and detail of a heating system. Those who live in flats, in gasless areas, or so isolated as to be without electricity, have special limitations. The infirm need effortless systems, and people who are often away from home need systems which run themselves most reliably. In at least one such case, the absence of gas, bottled gas can always be had and is an adequate substitute with a limited but sufficient range of appliances.

The very first choice is this. Shall we have a wet system? A dry system? Or (inevitable 25 years ago, unthinkable 5 years ago) single appliances with no linking system? The resurgence of the last calls for an explanation. In part it is because the choice of single appliances, and their improved technical qualities, have made them more competitive. In part the improvements in insulation standards have reduced heating requirements. But a large part of the change comes from the realisation that when the source of a heating system (usually a boiler) is working below its normal capacity, which for a good part of the time it probably is, its efficiency is not what it should be. We shall be looking at this aspect in the next chapter. Meantime we must decide between wet and dry systems.

This is a wide open choice in houses not yet built, into which the ducting for a warm air system can be built. For existing houses the choice is limited, as we shall see in Chapter 10, and we can turn our attention to wet systems. There is so much that can be said that the intention is to waste no time on types of system which are no longer recommended, such as gravity circulation. (If you own a country cottage with no electricity, there are plenty of older books which describe gravity systems.)

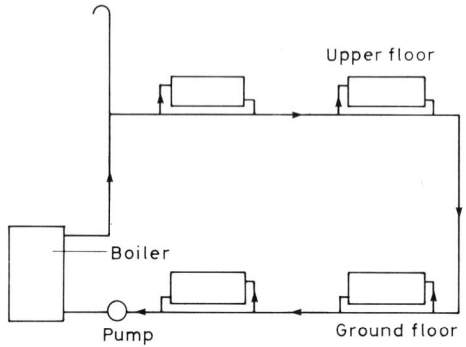

Figure 6.1. A typical single-pipe system

The first of the pump-circulated systems is that called the single pipe. It is in the small-bore range, i.e. the pipes used are in the range ½ in, ¾ in and 1 in if threaded iron, or 15 mm, 22 mm and 28 mm if copper or other. As *Figure 6.1* shows, the system consists of a single pipe which leaves the boiler and eventually, after serving whatever heat emitters are on its route, it returns to the boiler. On larger systems it

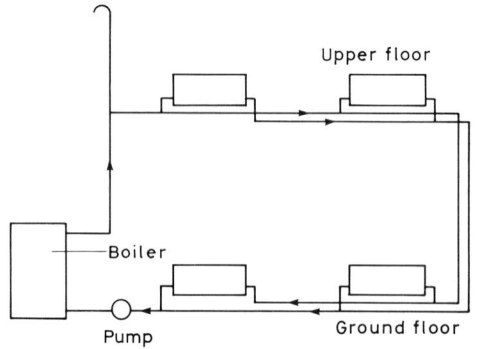

Figure 6.2. A typical two-pipe system

may be branched into zones without affecting the principle. Its popularity is diminishing because of the way in which the temperature becomes diluted on the way, each succeeding radiator returning cooled water to join the once boiler-hot water. Let us disregard it in favour of the two-pipe system, which has one pipe to deliver hot water and another to take all the returns, so doing away with dilution. There is a second advantage, analogous to series and parallel electrical connections. You need a two-pipe system with certain emitters, such as fan convectors, whose internal resistance is high and cumulatively would defeat the pump.

It is, then, the two-pipe system for us, and this will be described in Chapter 8, along with that other front-runner, Microbore.

We have not mentioned gas central heating, or oil ditto, for the very good reason that it does not enter into choices. You have chosen a system, for instance a two-pipe wet system. Now you will be able to decide which fuel to have, and here is a little guidance.

Solid fuel

Once this, like single appliances, was almost written off. But like them it is strongly on the way back, and like them it has developed. The Clean Air Acts have seen to it that not everything black may be burned, if it will create smoke. Hence there is a range of smokeless fuels, coke, briquettes, semi-coals like Coalite, anthracite and steam coal. They are not always all available 'off the shelf'. So begin by deciding upon a likely appliance. Then read the instructions to find out what fuel the makers recommend – not only type but size – and then approach local fuel merchants to see whether they can or will supply. We have no wish to exaggerate possible difficulties, but nothing could be worse than installing an anthracite-burning boiler only to be told 'We can't get anthracite up here'. In cases of real difficulty we would recommend complaining to the NCB at Hobart House.

Oil fuel

Oil does not have the consolation of an expected improvement, and before installing an oil-burning appliance we recommend making sure that the local supplier will guarantee, so far as he can, continuity of supply. Discuss also the best economic size for an oil storage tank. Larger tanks cost more, but often oil is sold on quantity discount. In any one case one could determine whether a saving on oil might justify spending more on the tank.

The local authority will put certain conditions on tank installation, its position relative to the house and to boundaries, provision of a fire valve, etc. Although disguise is not easy, do not bury a tank in a pit. It is more difficult to maintain against rust, and introduces the need to pump-lift the oil to the burner. See *Figure 6.3 (a)* and *(b)*.

Fuel oil offers a choice, between what is commonly called gas oil, and kerosine. These are also known as 35 sec. and 28 sec. oil, or Class D and Class C. Gas oil is marginally cheaper, but if you intend to have the appliance indoors you should choose the other. If the appliance is to go in an outhouse, then most pressure jet burners can be adjusted for either type of oil, but vaporising burners usually require kerosine.

Gas

Gas offers the greatest choice of appliances, shown in Chapter 9. Also there is no reason to suppose, on current evidence, that the availability of gas will be restricted for the next few years. Appliances bought for bottled gas can be converted if a mains supply is connected in the future. For various reasons gas is the popular fuel at the moment and it is well organised as an industry to give technical and other advice and to oversee the safety aspects of gas usage. People do not make sufficient use of the services which are available at gas showrooms, sometimes by showing a little persistence. Gas is the most versatile of the fuels, being available for

68

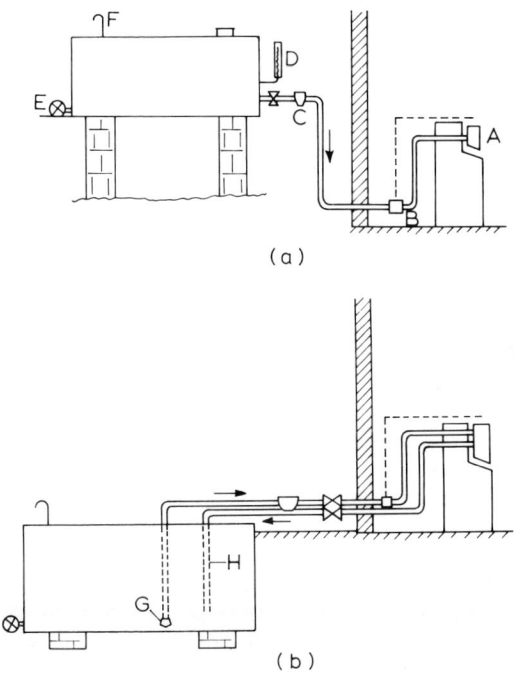

Figure 6.3. (a) Gravity oil feed via single pipe and (b) Suction oil feed via two-pipe system with continuous pumping. (A) Pressure-jet burner. (B) Fire valve. (C) Micronic filter. (D) Level gauge. (E) Sludge-drain valve. (F) Vent pipe. (G) Foot valve. (H) Return pipe

boilers, domestic water heaters, wall-mounted and free convectors, radiant room heaters, air heaters – the whole range of domestic appliances.

We cannot go on to sizing appliances until we decide how we shall run our system. A system running flat out all the time needs a bigger boiler than one which heats only a room at a time. Choice is simplified by describing ways of operating as follows.

● *Full central heating* is designed on the assumption that at any time every heat-emitting appliance may be at work, and that the house will be maintained at the chosen comfort temperatures when the outdoor temperature is −1°C.

• *Background heating* is similar to full heating *except* that the ambient temperature maintained will be enough to keep the chill off, leaving local areas to be boosted by other means (gas fire for instance).

• *Partial heating* resembles full heating but in this case it is not the whole house but only certain rooms which are equipped for heating.

• *Partial background heating*, which is self-explanatory.

It will be seen that the above categories range from the extravagance of full heating to the self-denial aspect of partial heating or the often barely comfortable background heating. But there is another way, which avoids those criticisms.

• *Selective heating*. Few homes have all their rooms occupied simultaneously. It is possible to draw up a shortlist of rooms which could very likely be occupied in this way. If there are two such lists, e.g. one by day, one by night, choose the one which design calculations show to have the greater heat demand. That is the total heat requirement.

It follows, then, that any other combination of rooms not of greater heat capacity could be satisfied by any arrangement made.

The method is to equip all rooms with appropriate heat emitters, such as radiators, making sure that they can be controlled on/off. Then choose the boiler or other heat generator to be large enough to supply only the shortlisted rooms. This makes a net saving on the boiler cost, as well as giving it a better chance to work at high efficiency. The basis should be that of full heating in temperature terms. On a few days in each year a very low outdoor temperature or an influx of guests might catch you out. But the economies are worth such rare upsets.

Design calculations

We referred to design calculations, and now is the time to see what they are. Taking a room at a time, they decide how much heat will have to be put into the room to replace what is lost, at −1°C outdoors and the design room temperature indoors. This they do by measuring the room surfaces and deciding what their U-values are. See Chapter 1, for U-values. Before going into the calculation we must settle what design room temperatures are, and you can insist that they are what you say. But with economy in mind, it should be remembered that overheating is an important factor in encouraging heat loss, therefore waste and increased cost.

The following list is worth considering, since we know that for most people it is about right.

Living room(s) 21–23°C
Bedrooms 12–18°C
Kitchen 18°C
Dining room 18–21°C
Bathroom 23°C
Hall 16°C

In passing it is worth noting the way in which heat loss depends upon the difference in temperatures between one side and the other of a wall, window, ceiling or floor. If the two sides are at the same temperature, no heat will pass. This means that from a thermal point of view, vertical and horizontal flats, terrace houses and (best of all) the old back-to-back terraces are far more desirable than the more trendy detached properties. Given reasonable sound insulation they are also socially acceptable.

Looking at another aspect of the same phenomenon, temperature is not the only determining factor. A wall subjected to much wind and driven rain cools much more rapidly than a sheltered wall at the same temperature. Indoor condensation is worst on exposed walls, and everything points to the need to shield exposed walls from the worst of the weather, for instance by tile hanging.

The heat loss calculation does not take exposure into account, which is in any case not easy to measure. Nor does it allow for a degree of habitation, say three people, and extra heat gain from electric light and television. We would treat those factors, and the occasional solar gain through windows, as a bonus, for heating can always be turned down but cannot be stretched beyond its maximum.

We have discussed design room temperatures, and the design temperature difference, which is the first plus 1°C, so now let us look at the other simple definitions. External wall area is the length and height of the total wall area which has outdoors on the other side, *less* the area of windows and any door to outside. Window area is the total if more than one. Floor area needs no explanation. Ventilation is the allowance for the number of air changes per hour (no.), bringing in room volume.

Then heat losses are:

External walls area A × U-value × temperature difference = watts

Windows area B × U-value × temperature difference = watts

Floor area C × U-value × temperature difference = watts

Ventilation $\dfrac{\text{room vol.} \times \text{no.}}{3600}$ × 1200 × temperature difference = watts

Add all those watts together, divide by a thousand for kilowatts, and you have the total heat loss for just one room. Repeat the exercise for every room you intend to heat. Then, if you have decided upon selective heating, add together the room totals of your shortlist of rooms, e.g. hall, dining room and living room. Then, if it is a boiler system and will be responsible for the domestic hot water, add another 3 kW for that. And that is the size of the heat generator you need.

You will need all the individual totals in order to choose emitters, e.g. radiators for the rooms. There is still a lot of support in the trade for adding a 'margin', which is supposed to cover contingencies and can be as much as 30 per cent. This can be useful for more rapid warm-up, or as a hedge against sub-zero temperatures, but for most of the time it is just another factor in increasing heat-generator rest periods and therefore adding to overall working inefficiency.

When you have worked out your heat requirements with a fair amount of accuracy (which doesn't mean to the nearest watt) and then find that you cannot buy an exactly matching piece of apparatus, you have a choice. You may choose to buy a larger or a smaller one than you need. It is rarely prudent to buy a smaller one, and so you choose the larger, shopping round if need be to find a unit which is nearest to what you want. Depending upon the relationship between what you want and what you get, your apparatus will have an unavoidable amount of oversizing. This may be counted as a 'margin' and deliberately to add another one is quite unjustified.

7

Solo heaters

Middle-class houses of the Victorian and Edwardian periods can be recognised by the large number of chimney pots they carry, indicating the proliferation of coal fires indoors. Modern owners have hastened to block them off, with or without capping. Nobody wanted crude coal fires, everybody wanted central heating, and 25 years ago that represented the choice. Now the pendulum has swung back again, and the near-monopoly of central heating is being challenged by a new generation of single-room heaters, and anyone thinking about heating should take heed of the fact and its attractive features.

Some choice has existed for years, of course, but it was probably the Clean Air Acts, bringing about the virtual death of old-style open coal fires, which started a strong movement of revival. This movement has justified itself by the amount of innovation and development which has gone into the modern appliances.

Electrical appliances have to be credited with about 100 per cent efficiency for the simple logical reason that whatever energy is fed in is delivered as heat on the spot – there are no flue losses, for instance. For all the rest we may note the existence of two efficiencies. There is the 'rated' efficiency, that obtained under test, including approval tests. And then there is the day-to-day figure which derives from all the ups and downs, stops and starts and load variations which are part of normal running. Although the second figure is less than the first, the difference is in the main less for individual appliances than it is for a central plant, particularly a boiler. One reason for this is that single appliances tend to be run at or near their maximum or rated output, whereas boilers are frequently underloaded, and subjected to considerable Off periods, when their mass cools down without doing any useful work. (That is why we detest the practice of oversizing boilers, including deliberately adding 'margins'.)

To the likelihood of better efficiency we can add a possible economy, namely heating only one room when only one room is needed, instead of having other emitters at work off the system; and in terms of cost and domestic upheaval there can be quite an advantage at the start.

Electrical appliances

We begin our survey with electricity which, if it were not for the high price of normal tariff current, might eliminate its competitors. As it is, it owes its competitive position to its great convenience. Almost without exception, electrical appliances may go anywhere. They need no flue, no special wall inside or out, no pipework; just a cable or an available power point.

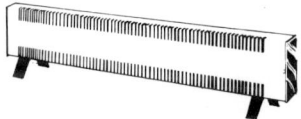

Figure 7.1. Electric skirting heater. Some are attached to skirting boards; others, like this, have feet (courtesy Dimplex)

A quick look across the field shows wall-mounted and floor-standing versions of radiant fires; radiant convector fires; skirting heaters (*Figure 7.1*); radiators resembling hot-water radiators; fan convectors; natural convectors; towel rails; infra-red heaters. Some are portable; others, including the wall-mounted ones, are to be permanently wired in, in accordance with IEE Regulations. The usual reservations apply to their use in bathrooms, and this is always spelled out in the instructions. Otherwise the purchaser has very little to be concerned about. In use there are certain commonsense rules. Fabrics and other combustibles should never be hung over a fire which is out, if there is a chance that it will come on – by clock control, for instance – when nobody is about.

Storage heaters
Electric storage radiators – the common storage heaters – have for years been classed as central-heating appliances, which they are not. Now they can be reclassified without loss of prestige. Their outstanding advantage, to which they owe their existence, is that they run on cheaper off-peak current. Despite desperate attempts by the competition to show otherwise, this gives them a running cost in a comparable bracket with other forms of central heating, after one has allowed for fundamental differences in the manner of heating.

Storage heaters are overall the cheapest system to install, since apart from their purchase price the only cost is for wiring. Each heater must be wired right back to the board and into a consumer unit (fuse box) given over to off-peak units only and controlled by the clock which the Electricity Board puts in. It is usual, convenient and economical to have the immersion heater in the hot water cylinder wired into the off-peak. But unless the cylinder is large enough to hold a day's supply of hot water there must be auxiliary switching for peak-load current as well.

Storage heaters fall into two classes. The first is filled with heat-storage material, usually silica brick or block, covered in insulation and heated by electric elements let into grooves in the blocks. The insulation is designed to allow a calculated heat leak, and the heat output of the unit rises to a peak some time after heating ends, and then tails off until the next charge is due. The mid-day input boost used to arrest the downward slope of heat output, but since that has gone the plain heater has lost some of its popularity. This is in fact the type which is sold through local papers for between £5 and £10. For those whose chief need for warmth is in the morning, these are real bargains.

The second class of heater is more controllable. Natural heat leak is reduced by increasing the insulation. Then, the

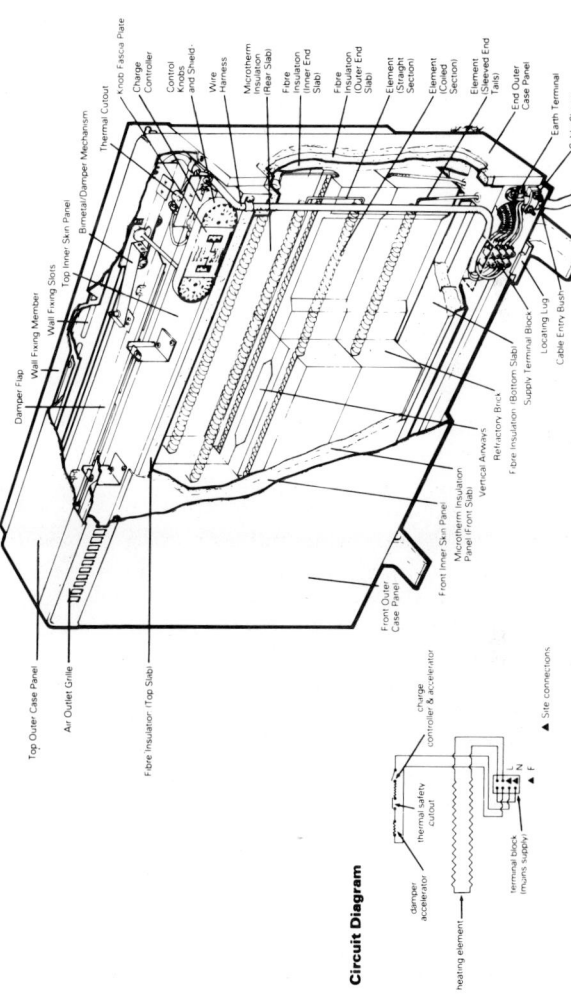

Circuit Diagram

Figure 7.2. Construction detail of the Dimplex SC range of storage radiators. Note the flap damper at the top of the vertical airways (courtesy Dimplex)

Figure 7.3. The Dimplex XT range of storage radiators are only 150 mm deep (courtesy Dimplex)

filling has a hollow core or cores up which air can pass and become heated. But the passage of air is subject to damper control, which can be seen in *Figure 7.2*. The damper itself may be manually operated, or increasingly is automatic so that opening, hence room warming, can be pre-planned.

This type of unit enables a substantial amount of warmth to be kept back for, say, the evening.

It is obvious that, while the heater is on an off-peak line, any control gear is bound to be on the day rate. This calls for special care in wiring by anyone who has a three-phase supply led in. To avoid any chance of trouble, both supplies should be taken from off the same phase at the board. If there is any doubt about this, you should consult the Electricity Board.

Storage heaters were for a long time widely known for being bulky, and were often rejected on that account. Much development work has gone into slimming them, so that now they need take no more room than a double-panel hot water radiator, as *Figure 7.3* shows.

No special floor covering is needed. Safe stability can be assured by fixing the heater back to the wall where it stands, so long as the model does not have rear air discharge. For every type of unit there is one very important warning. Do not allow objects, in particular clothing and textiles, to stand on the heater, since this could lead to a blown (melted) safety link and the heater out of action. If there is great need for the airing function, it is often possible to buy a shelf, with brackets which hold this a safe distance above the heater top. Although storage heaters may be fixed beneath windows, by the same reasoning which applies to hot water radiators, curtain bottoms must be kept well above the heater top.

A fairly recent development with storage heaters is input control, designed to sense trends in the weather and to modify the input accordingly. It is claimed that this can make considerable savings, in which case it must be more reliable than what it replaces. Input control we have always had, manually guessed at by the householder, who will probably be relieved to lose the job. *Figure 7.4* shows a few results of varying the relationship of input and output controllers.

A brief note about off-peak current, which used to be an optional amount varying from some 7 to 11 hours, the latter including a mid-day boost already mentioned. That has all gone, and we are left now with a flat 7 hours, known at the present time as the Economy 7, which runs from either

79

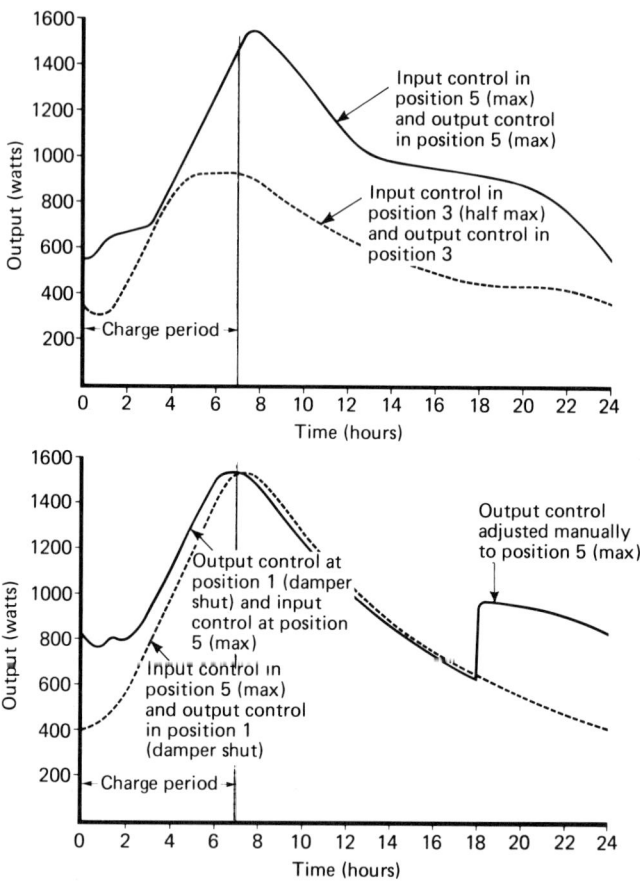

Figure 7.4. Some interesting performance graphs from Dimplex, showing how their SC 24 storage radiator output behaves in varying conditions of input and output control (courtesy Dimplex)

midnight to 7 a.m. or half an hour later than that. There is, however, another difference.

With the present tariff, *all* you use between the stated hours is at the cheaper rate. You may therefore gamble with the Electricity Board that you can really save by adopting their tariff. It all depends how much of your normal usage you can divert to the small hours. The washing machine and the electric cooker, both heavy users, can be automatically programmed. Lawn mowing is not among the likely uses to benefit, but it is worth thinking about what might be transferred to the night shift.

Oil-fired appliances

Looking briefly at what oil-fired appliances have to offer, we have already made it plain that free-standing flueless paraffin stoves are not even considered. And then, it is noticeable that whereas some ten to 15 years ago hearth-mounted oil heaters, with or without back boiler and most of them

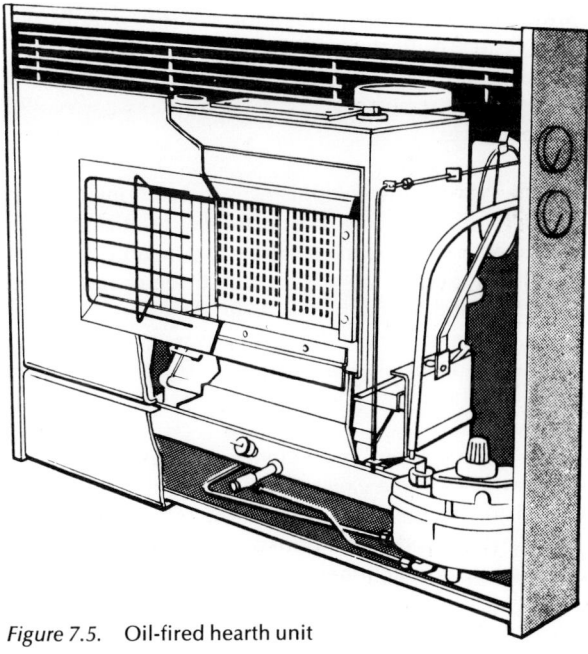

Figure 7.5. Oil-fired hearth unit

Scandinavian, were advertised everywhere, nowadays they are almost unheard of. They were, in the main, free-flow convectors, generating a lot of hot air in one room and allowing it to drift through the house. In spite of some fantastic claims for temperature levels, many people seem to have been satisfied.

The best of these, which happens to be British, resembles a very large gas fire (see *Figure 7.5*). For instance, it fits into a hearth, and has a radiant section similar to that of a gas fire. Behind this is the convector section, with or without back boiler, and this generates a large quantity of warm air. It is usual to fit a fan at high level in the wall of the room, so that much of the warmed air can be sent on its way, usually into the hall, whence it can go to other rooms and upstairs. A unit of this kind can have its fuel storage built into the appliance, or it can be fed by pipe from an outdoor storage tank. With the back boiler it is a very civilised amenity for the isolated house which has no gas and which for any reason does not want a full system of heating.

Solid-fuel appliances

The rôle of solid fuel in solo appliances is traditional, and in the past such appliances were disastrously inefficient. The average efficiency of an open coal fire was probably below 10 per cent. The modern open fire, though three or four times better than that, still has a long way to go, and the incorporation of a back boiler helps it along the way. But too much must not be made of simple efficiency, for a great many people find great comfort and satisfaction in the movement of real flames.

The great improvement in efficiency of open-fronted appliances has quite simple causes. Principally, the amount of air admitted to the fuel bed from below is rationed by an adjustable damper; and the amount of room air which can be whisked away up the chimney is limited by restricting the throat into the flue, and this is improved further by using a flue liner as well.

82

This shows that some improvement is within the capability of people using ordinary materials. But correct design still counts for a lot. Moving on from the open fire we find appliances which are half way between an open fire and something quite hidden like a boiler. These are the closed or closeable room heaters. Efficiency is much greater with the appliance's door closed, and so a window is provided, to allow the user to keep in touch with events. Appliances of this kind are wholly contained within a cast-iron casing. They may be inset or free-standing, in the latter case being set on the hearth in front of a chimney and connected to the chimney by a flue pipe.

They may be front-fed or hopper-fed, and they may be complete with back boiler, or without one. And of course they come in a variety of sizes, physically and thermally speaking. As an example, look at the Parkray 99G (*Figure 7.7*). Every appliance bought is supplied with full instructions, from which we take only a few points.

The Parkray 99G, as can be seen, has a back boiler, for which there must be space at the back of the hearth. A fully indirect cylinder must be used, along with a radiator in parallel with the primaries as a heat leak. Once the unit is in position and level it is anchored to the floor of the hearth, after which pipe connections may be made to the boiler, through an excavation in the side wall of the chimney breast. The water system is then completed and tested for leaks before proceeding with the heater.

Run a short length of 150 mm flue-forming pipe, to reach from the heater spigot into the chimney (assuming no liner). Check first that the chimney is quite clean and sound. Make good the space around the flue-former and at the other excavation, filling with vermiculite concrete. There must be no air leaks in these areas.

Air leakage is also prevented around the rim of the heater body by a sealing rope which butts up against the hearth front.

The efforts so far have resulted in a finished boiler, and a secure chassis for the heater section. This is now entered into the chassis, firm and level, and secured by the fasteners

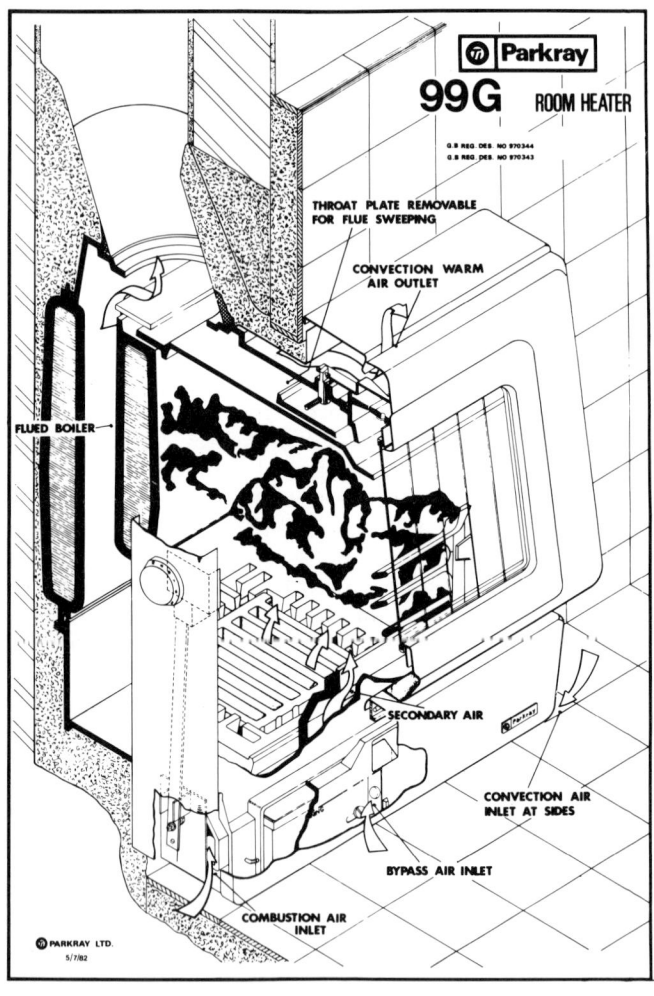

Parkray

🔘 Parkray

99G ROOM HEATER

G.B REG. DES. NO 970344
G.B REG. DES. NO 970343

THROAT PLATE REMOVABLE
FOR FLUE SWEEPING

CONVECTION WARM
AIR OUTLET

FLUED BOILER

SECONDARY AIR

CONVECTION AIR
INLET AT SIDES

BYPASS AIR INLET

COMBUSTION AIR
INLET

🔘 PARKRAY LTD.
5/7/82

Figure 7.6. Section view of the Parkray room heater model 99G showing the travel of convection air and the relationship of back boiler and fire bed. This unit is mainly inset

supplied. Many details have to be checked at assembly stage. The thermostat must work and the control knob rotate freely. The grate-shaker mechanism must be free to shake. The fire door must close fully, or else be adjusted so that the maximum gap on closing is 0.008 in. When these and more matters have been attended to, the appliance is ready for firing, using paper and wood, or firelighters, or a gas poker. Do not run to maximum for at least 24 hours.

The operation of the control thermostat is fully explained, including its use when setting for overnight burning or 'slumbering'. There are notes on ash removal, advice to remove only cold ash, and to prevent an accumulation which could touch the grate. There is a damper control which will determine what proportion of the total heat goes into water.

Recommended fuels are Anthracite stove nuts; Coalite; Phurnacite; Rexco Royal; Rexco Superbrite; Sunbrite doubles; and Welsh Dry steam coal nuts.

Performance of the 99G with the boiler damper closed is given as space heating 2.3 kW (8000 Btu/h) and boiler output 8.7 kW (29 500 Btu/h). With the boiler damper open we get for space heating 2.2 kW (7500 Btu/h) and for boiler output 9.4 kW (32 000 Btu/h). From those figures it is clear that the 99G is intended for, and has a large capacity for, running a wet system of radiators.

Operation of this type of appliance is simple, but not entirely positive on account of variations in fuel, and a user therefore might need to find after short experience such things as the best setting for night burning. But there is one matter which cannot be stressed too often. Avoid air leaks. On front-loading units shut the door tightly on a clean frame. On hopper-fed models it is even more important to see that the hopper lid is tightly shut. Any appreciable draught through the hopper could cause it to take fire. But always, local overheating and loss of efficiency are the risks.

The Baxi Burnall

The broad description given above will hold good for most modern heaters, and details special to any one make will be found in the instructions supplied. All models which have

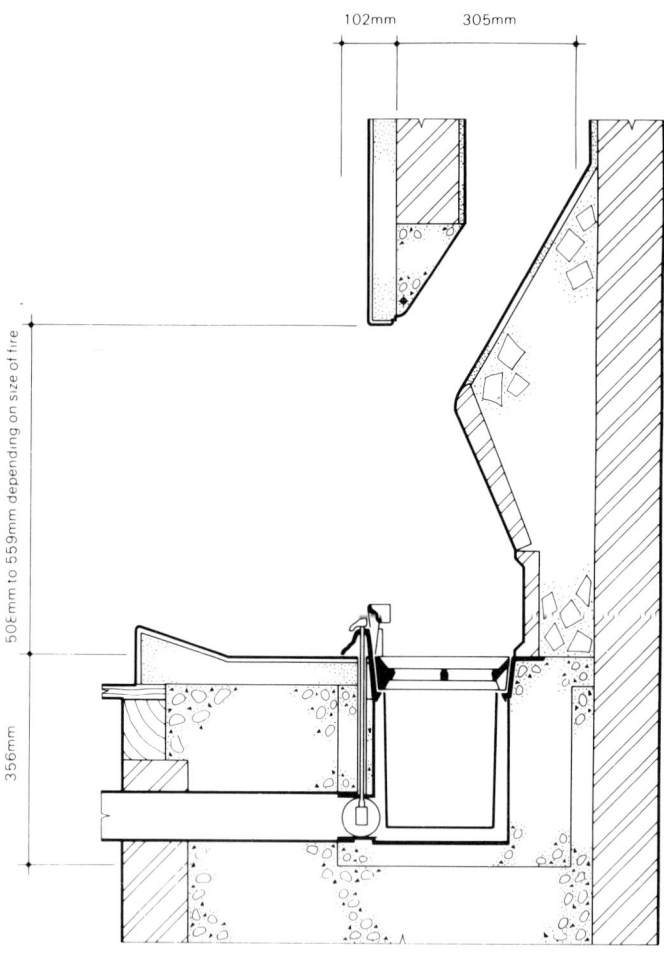

102mm   305mm

506mm to 559mm depending on size of fire

356mm

Figure 7.7. General arrangement of under-floor draught solid-fuel fire. Note the draught regulator (courtesy Baxi Heating)

received official approval from the Domestic Solid Fuel Appliance Approval Scheme (DSFAAS) have their instructions approved in the process, and of course no one would buy an appliance not approved. But there is one unit which is out on its own, and has been for a number of years, building up a solid body of support. It is the Baxi Burnall, noted for the way in which it relies upon under-floor draught.

In Chapter 5 we referred to the possibility of supplying some combustion air from beneath a ground floor, or via a duct from an adjacent outdoor wall. Such an arrangement is an essential part of the Baxi. There are two parts to the package. One is the draught arrangement. The other concerns ash removal. It is a part of the technique that the fire, to take full advantage of the under-draught, has its bars slightly below normal floor level. A conventional dog grate is raised a little so that ash accumulates beneath but still above floor level. In the Baxi, ash is already below floor level and can only go lower. Hence Baxi has a sunken ashbox, let into a prepared pit. The box is able to hold three days' make. After that it may be lifted straight out; or, in the case of outside wall chimneys, removed from outside through a prepared aperture; or, using twin rotary ash boxes, removed one box at a time leaving one in use. The choice has to be made during the structural arranging for installation. *Figure 7.8* shows the three methods of ash removal.

Regulation of air flow is by a damper, lever-operated. The method of air introduction, via a restricted channel, increases the velocity of the air and this enables the fire to burn some of the harder fuels successfully.

The makers warn that an excessive flow of air into the room through a constantly open door will prevent the Baxi fire from working. It will then work only if a forced draught fan (available) is used. On the other hand, while moderate draught proofing is approved, the room must not be sealed, or a smoking chimney will result, just as with conventional grates. But the use of a throat restrictor will cut down on the volume of draught which will be taken from the room.

This unit may be had with high- or low-capacity back boiler, or without.

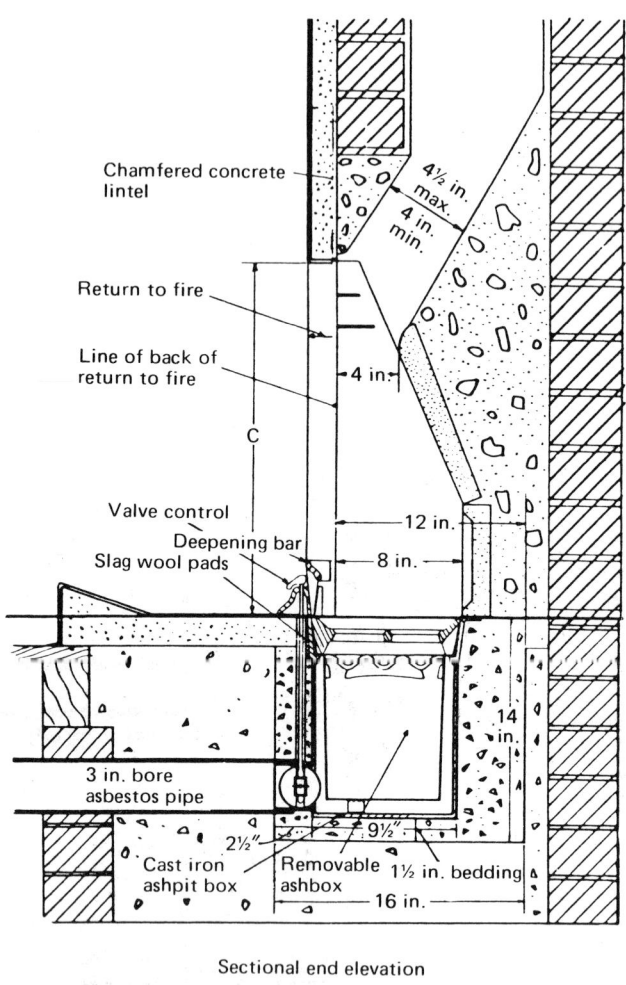

Chamfered concrete lintel

Return to fire

Line of back of return to fire

C

Valve control

Deepening bar

Slag wool pads

3 in. bore asbestos pipe

Cast iron ashpit box

Removable ashbox

4½ in. max.

4 in. min.

4 in.

12 in.

8 in.

14 in.

2½"

9½"

1½ in. bedding

16 in.

Sectional end elevation

(a)

Figure 7.8. Three standard methods of dealing with ash removal: (a) indoor removal of a single ash box; (b) indoor removal of one segment only of a rotary ashbox; and (c) outdoor removal of the ashbox (courtesy Baxi Heating)

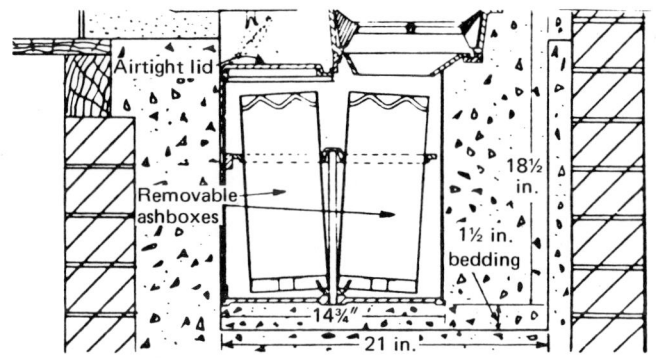

Sectional end elevation

(b)

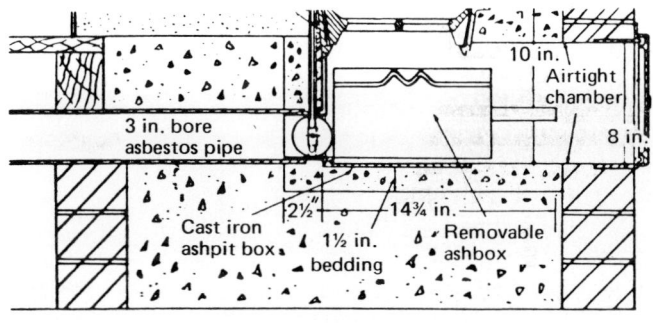

Sectional end elevation

(c)

Gas-fired appliances

We come last to gas-fired appliances, where the choice both of types and of manufacturers is very wide. To look across the range first, we can see:

Gas fires (radiant/convectors); back-boiler units, usually with a fronting gas fire. Gas fires may be wall- or hearth-mounted, and a small number can connect to a balanced flue, having a closed front. A recent survey showed 49 models currently available. If not all of these are obtainable from gas showrooms, try the builders' merchants.

The output from most gas fires falls within the range 3.0 to 4.0 kW (10 000 to 14 000 Btu/h) to suit most rooms. Output may be reduced easily by a simple turn of the knob to half gas, or to eliminate the outer radiants, or some such means. Lighting, these days, it usually by piezo-electric spark, which should last for ever, though at least one model has a pilot light.

Styling is important in items which are so prominent. The trend towards the 'wooden box' is giving way to more ornamented period pieces, not always an improvement, and indeed newly styled box surrounds are coming back. There is also a slight tendency to expand the case and invite other uses, mainly ornamental.

A move which is proving to be far more popular than it deserves is that towards ceramic log 'living flame' fires. By

Figure 7.9. The Baxi Bermuda W2, a piece of furniture described as a complete fireplace, with large back boiler (courtesy Baxi Heating)

these, the gas industry which worked so hard to improve the efficiency of gas fires has let it slip away again. Readers are urged to resist the temptation to acquire a 'living fire' if they value efficiency.

Radiation, as we pointed out elsewhere, is far more effective than warmed air in promoting a feeling of comfort, and the gas fire makes radiation easily and quickly available.

We must mention flueless convectors and open-flue convectors, but the convector market has by now gone largely to the balanced-flue convector, known also as the wall heater or unit heater. Flueless heaters have been mentioned as a source of condensation, and a limit is placed upon their size. This maximum is 50 W per m^3 of room volume, or 100 W in a more open location. A flueless heater can be useful in such conditions as a hall with stairs off leading to a landing i.e. a fair amount of volume, when the hall has no outside wall to accept a flue.

Open-flued convectors follow the same rules, for installation etc, as gas fires. They are in effect fires without the radiant section, which is useful where the very young or very old have access. It is useful too in libraries and museums and the like, where general warmth is needed but the localised heat of radiation would be unwelcome.

The same advantages apply to the balanced-flue convectors. It is these which have put convectors on the map, partly no doubt because with a balanced flue they are very versatile. One must not overlook too that they have a high efficiency, as well as simplicity. In their standard form they need no electricity, and although this matter is receiving attention in some quarters, we must respect the view of a major manufacturer that the inability to fit a clock and other electrical gadgets is far more upsetting to a few pundits than it is to the public. The ultimate in simplicity and cost efficiency must surely be with Baxi, whose unit dispenses with a bypass – always a grudged user of gas – and has its main burner lit direct by a piezo igniter.

At the other end of the scale is Drugasar who have devised an add-on system for applying time etc control for up to 16 units per controller.

Installing gas fires

The installation of gas fires has certain features common to all, and of great importance. These are aimed at achieving correct combustion, and correct disposal of the products of combustion – which is *not* out into the room. Let us take the common case of a fire being fitted into an existing hearth. We will assume that the chimney has been cleaned and inspected, and the hearth cleared out to give ample room for any later falls of solid matter. Then fit the closure plate. This is a sheet, commonly of 22 swg half-hard aluminium or asbestos millboard, which is large enough to fit over the entire hearth opening. The first thing to do with this is to measure it against the fire and mark and cut out an aperture for the fire's flue spigot. A clearance of about 3 mm (⅛ in) should be left, partly to allow 'relief'. This is in addition to the slot of about 150 × 50 mm (6 × 2 in) which is cut at the bottom as shown in *Figure 7.10(a)*. This relief has two purposes. It

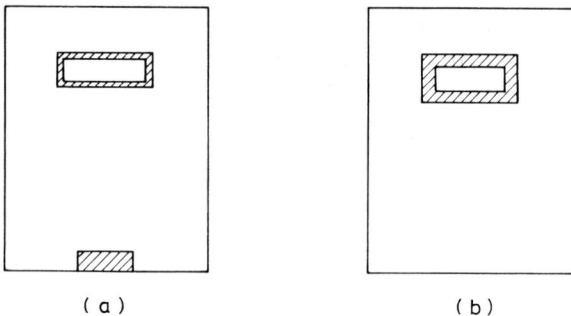

(a)   (b)

Figure 7.10. Cutting the closure plate (a) with modest clearance around the flue spigot plus bottom edge relief and (b) more generous relief around spigot and none at bottom

allows some room air to escape, to promote room ventilation. And in the event of a sudden downdraught it does provide relief, safeguarding the flame. It is permitted, though not so useful, to omit the bottom slot and add the area to the excess around the flue spigot (*Figure 7.10b*). Having cut out the holes, one may now fit the closure plate. This is fastened to the hearth front, using a heatproof sealing tape, usually adhesive and sealing all four edges. It is most important that the closure plate shall be an airtight fit.

When that is in place, check at the spigot opening with a smoke match to ensure that there is a constant up-draught. If there is neither up- nor downdraught, try to start the chimney by inserting lighted paper. If there is still no result, do not fit the fire but re-examine the chimney for blockage.

Once that is settled the fire may be eased into place, the flue spigot entering the closure plate to its full extent. Secure to the wall if a wall-mounted fire, and make the gas connection.

The next serious business comes when putting the fire to work. A typical fire provides two routes for its flue gases. The intended one is more difficult and relies upon chimney pull to achieve it. If this, which is usually around 0.1 mbar, is below that level, the flue gases might take the easier route out via the canopy and into the room. This is called spillage. Advice should be sought, but a naturally deficient chimney might need a liner fitted since this will increase flue-gas velocity.

The effects of too much draught are just as much in need of attention. They call for a throttle, which can be had in the form of a restricted spigot. But this is an entire area deserving of professional attention and instrumentation.

If a gas fire is fitted into a precast concrete flue there could be a risk of local overheating because of the narrow section of the flue. To avoid this, a so-called cooler plate is fitted to the back of the closure plate. This is a metal baffle standing off from the spigot outlet which intercepts and deflects the flow of hot flue gases over a wider area.

Information about this may be had from the flue or fire-maker or from the gas showroom.

A floor-standing gas fire must stand on a noncombustible material, usually the hearth itself, but metal, marble or similar will be acceptable. No such provisions apply to a wall-mounted fire, which may even be over carpet, provided only that the top of the burner is not less than 230 mm (9 in) above the floor.

At any time subsequent to installation when the fire is in any way disturbed, for example in servicing, the seal around the closure plate must be checked and if necessary made good.

Installing convectors

The installation of open-flued convectors is in all respects similar to that of gas fires.

For balanced-flue convectors it is an exaggeration to say that they may go anywhere on an outside wall. There are in fact two limiting factors. The first is the siting of the balanced-flue terminal (see Chapter 5). The second factor applies indoors. There are certain reasons why the heater needs space around it: for servicing; for satisfactory circulation of air to be warmed; to avoid too close contact with curtains; to be in a good position to spread maximum warmth. Most of these are set out by the maker, whose instructions should be followed.

Once a position has been chosen, the first job is to mark off the position of the flue terminal and the places for wall-securing bolts or screws. The second job is to prepare for and fit the terminal. An oversized hole is needed, and dimensions are usually given. Although the standard wall is now 11 in (28 mm) there are plenty of others, solid brick at 9 in (23 mm) or 16 in (40 mm), stone walls of any dimension, timber walls quite thin. In consequence a popular type of terminal is telescopic, and so can be adjusted to fit. Occasionally one has a cutting length instead. If you are faced with a situation in which the terminal supplied is not long enough, you should not assume that extrapolation is quite safe. Consult the maker.

Once the terminal is in position, a good deal of the surplus excavation may be made good, by cement etc. But the terminal should be separated from hard building material by a soft fireproof liner, e.g. asbestos cloth. In brick and stone walls this caters for expansion. In combustible walls it is an essential precaution, and its thickness is specified by the maker. There might be other requirements, which will be stated in the literature.

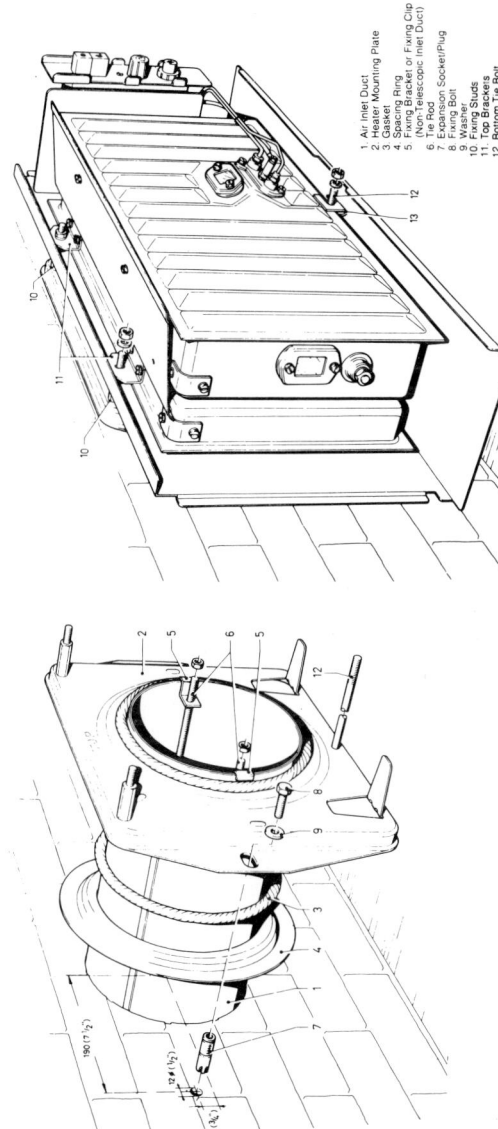

Figure 7.11. These two illustrations show the sequence of installation of the Drugasar Horizon range of heaters. Points to note particularly are (a) the flue terminal supplied is suitable for wall thicknesses from 50 mm (2 in) to 600 mm (23 in) with cutting; (b) the heater-mounting plate is held to the wall by expander bolts; (c) the heater is secured to the mounting plate by studs at the top, and tie bolt at the bottom; (d) the gaskets must be in position and compressed during tightening; and (e) there is a right way up for the terminal. The top is marked TOP

1. Air Inlet Duct
2. Heater Mounting Plate
3. Gasket
4. Spacing Ring
5. Fixing Bracket or Fixing Clip (Non-Telescopic Inlet Duct)
6. Tie Rod
7. Expansion Socket/Plug
8. Fixing Bolt
9. Washer
10. Fixing Studs
11. Top Brackets
12. Bottom Tie Bolt
13. Bottom Clip

Offer the heater up to the terminal, and check the fastener positions. Take the greatest care in getting sound holes in the wall, to carry toggle bolts or wall plugs. Subsequent collapses are always traceable to inadequate fixing into walls.

When finally bringing heater and terminal together, take great care to ensure a tight seal between the heat exchanger (on the heater) and the flue gas connection (on the terminal). There will be no later opportunity to work on this joint, and perfect room sealing is a condition of satisfactory operation.

Once the heater is firmly in place, gas and, if appropriate, electricity may be run to it. During servicing, check once more that the heater is rigidly attached to the wall.

8

Wet systems

As a result of following Chapter 6, we have decided some basic facts.

1. We know how many rooms, and which ones, will be heated.
2. We know the design heat loss for each room, and so can choose a suitable radiator or convector for each.
3. We have decided that our wet system will also supply domestic hot water (and if not we will simply ignore all future reference to that aspect of the installation).
4. From the total of design heat losses we know what size of boiler we want – in selective heating the total of the shortlist.

Location of equipment

We can now begin to make physical plans for creating a system, which means placing the various bits of equipment in their appointed places within the framework of the house, and then joining them all together with a circuit of pipework.

Since the domestic hot water side will be worked by gravity we will think of that first. Gravity is not a strong force, and to make sure of a good circulation we must avoid long horizontal pipe runs. This means that the boiler and the hot water cylinder should be in a close vertical relationship. Usually the cylinder is in the bathroom, so the boiler should be as nearly as possible vertically below the bathroom. Of course, if there

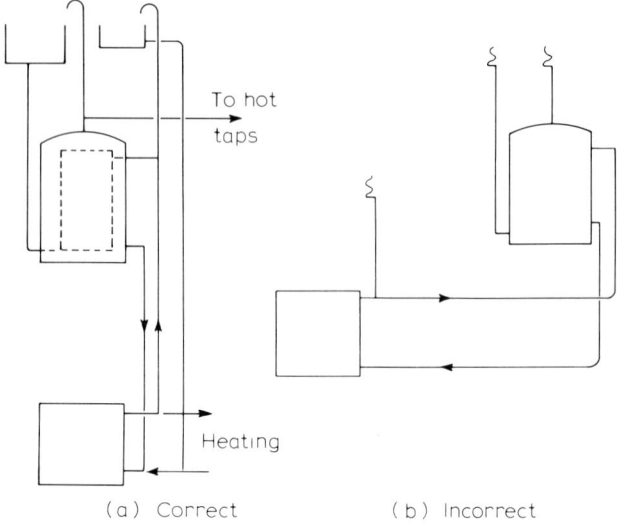

To hot taps

Heating

(a) Correct   (b) Incorrect

Figure 8.1. The boiler and hot water cylinder need to be in a close vertical relationship (a), avoiding long horizontal pipe runs (b)

is a much more suitable place for the boiler then one must think about relocating the cylinder, e.g. in a bedroom cupboard.

However, remember that long pipe runs for hot water are very wasteful, and most hot water is required in the bathroom or in the kitchen, which usually is below the bathroom. Hence boilers are usually put in kitchens.

The next pieces of fixed equipment are the heat emitters: radiators, convectors, fan convectors and skirting heaters. Which type suits you best?

The safe norm is the *radiator,* not least because it gives up to half of its output as low-temperature radiation – see Chapter 10.

The *skirting heater* is a kind of dwarf radiator, which has several good features. It is unobtrusive (nobody would claim that radiators are objects of beauty) and it is remarkably good at promoting a low temperature gradient, i.e. the ceiling is

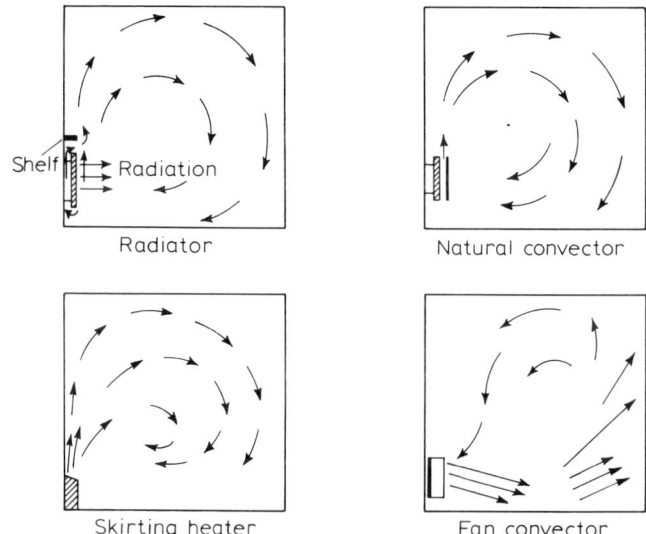

Figure 8.2. Four different ways in which a room is warmed

not made much warmer than the floor. Its lack of popularity in the UK has always been due to its higher price, while physically it does require the skirting, and no sideboards or other furniture in the way.

Convectors do not emit any radiant heat, and domestically are often preferred in households with very small children where high surface temperatures are unwelcome.

Fan convectors again have no radiant factor, but they do have a high degree of control, by thermostat and clock, to have heating quickly whenever wanted, in the right amount. They are not suited to bedrooms. Fan convectors must be situated in such a way as to give a clear path to the fan-driven warm air.

We have decided where fan convectors must go, and in irregular rooms they will be roughly half-way along a long wall. Skirting heaters usually have little choice, since a certain length has to be accommodated in order to get enough heat output. And natural convectors we can take as following radiators, since apart from radiant content they are similar. Meantime the placing of radiators has become a source of argument, often confusing.

Until double glazing came along it was widely accepted that radiators under windows produce an up-current of warm air which overcomes the natural descending cold air current from the glass. It is now being argued that this leads to a small but significant loss of heat. To which we reply that there are occasions when comfort comes before stark efficiency, and we continue to support the radiator-under-window position. But time is bringing a modification, in that double glazing is becoming more common. And where there is double glazing, the cold downdraught is mainly eliminated, so the position of radiators is open to choice.

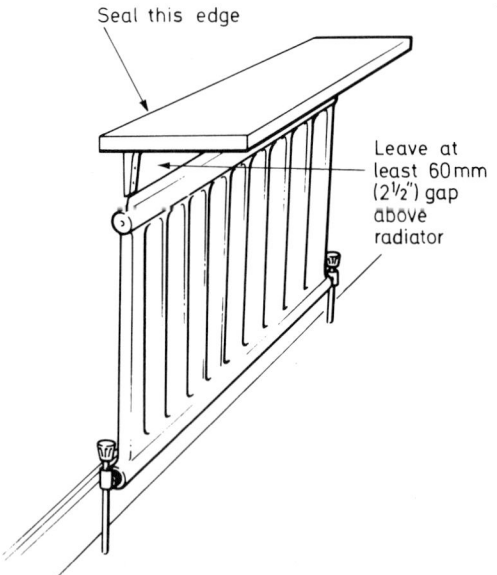

Figure 8.3. Radiators not beneath windows should have a shelf fitted to keep the wall clean. This should project about 50 mm (2 in) over the front and at each side, and be sealed to the wall

Two other controversial points about radiators deserve mention. First, it is claimed that radiators against outside walls cause more heat loss through the wall than if only warmed convected air played on the wall. This is true, and we intend to show that there is a positive virtue in putting radiators elsewhere.

The second point deals with the reflective back panel, a panel of polished metal which is to be fastened behind the radiator in order to save heat. If backed by an insulating sheet, or if an insulating sheet is used instead, it tends to save that small amount which escapes from radiators on outside walls. On inside walls it serves no purpose at all, since it is a good thing for warmth to be absorbed by the brickwork, to be given up later. Meantime the reflective sheet has the effect of reducing the total heat output from the radiator, which is rarely welcome.

Let us now think about radiator placing, where double glazing has given us a free choice. We must keep to the elementary rules, mainly that the radiation shall be able to 'see the room', not have heavy furniture in the way; also that the radiator, if only one, shall be in the main body of the room and not, for instance, in a minor L part. See *Figure 8.4.*

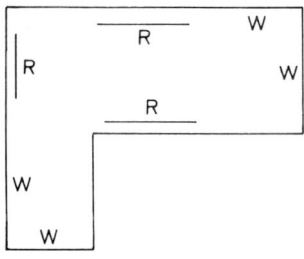

Figure 8.4. Right (R) and wrong (W) location for a radiator

Subject only to that, we can save money and work on pipe running, by choosing deliberately short runs off a simple out-and-home flow and return. *Figures 8.5* and *8.6* illustrate the point, and show how a further saving might be possible in some cases, by having back-to-back fixing.

The out-and-home method is ideally suited to tall narrow houses, where the flow and return can often be run up the

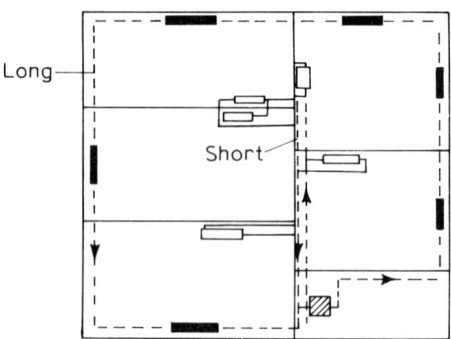

Figure 8.5. Long and short pipe runs

stair well; and to bungalows which are the horizontal version of that type, with rooms branching off a corridor. But scarcely any design of house or bungalow fails to offer a chance to design a minimum circuit, a sort of inner circle. Remember that the 'radiators-under-windows' scheme always uses the maximum amount of pipe, because windows are on the house perimeter.

Placing radiators in a correct vertical position is important too. They should be low down, to give maximum travel to the rising warm air current, and to let the radiant fraction work at the right height. But they must still be above the skirting line, to allow unrestricted access of air under and up behind them. The wall brackets supplied will hold them at the correct distance from a plain wall face. See *Figure 8.8.*

Measure bracket positions with the radiator held true and level. Drill the wall for plugs, and fasten the brackets with wood screws (usually supplied). Screw the inlet and outlet radiator valves, or radiator thermostat valve and stop valve, into the radiator before setting it on its brackets. Use a jointing compound lightly, on male threads only.

Piping

In Chapter 6 we gave the reasons for choosing a two-pipe system. The next choice is the material, and there can be little

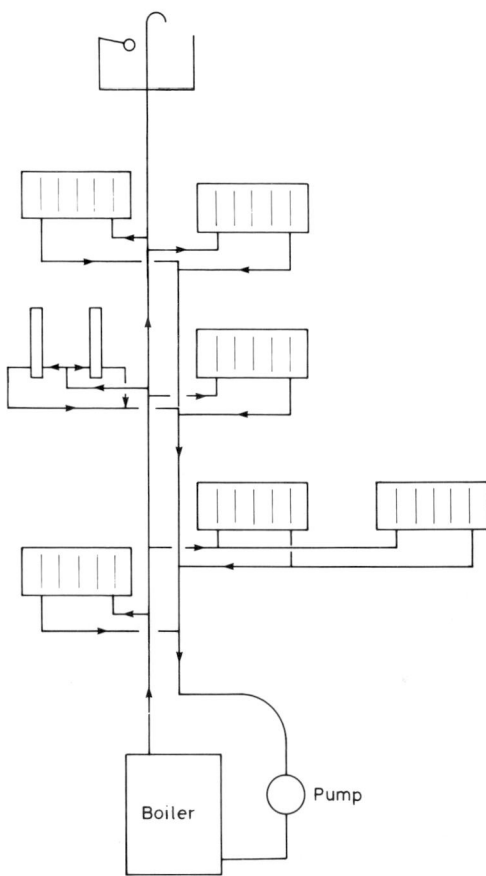

Figure 8.6. An economical pipe circuit

doubt that it will be copper. Plastics have at long last been officially admitted as suitable for both hot and cold water, but it will be a while before such pipes become widely available in precisely the correct type and formula.

Stainless steel is an adequate substitute for copper. But, to be realistic, where will you find it? Certainly not on the shelves of every builders' merchant who stocks copper.

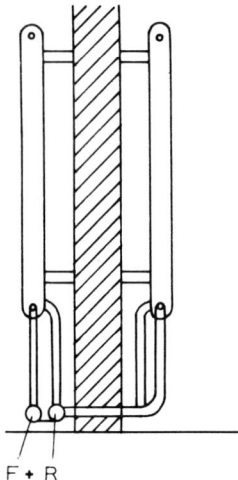

F + R

Figure 8.7.
Radiators back to back

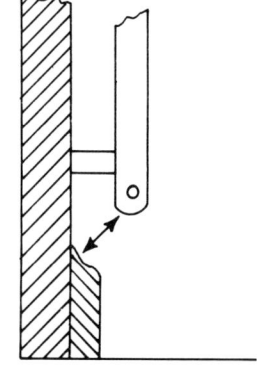

Figure 8.8.
Skirting clearance

Black iron is unlikely to be considered, not only because of its rust potential but because it is much less handy, for bending, cutting etc, than copper. And a galvanised pipe loses its protection as soon as it is sawn.

Next we want a plan or design for pipe runs. Start with the basic fact that everything must be joined up, in the sequence

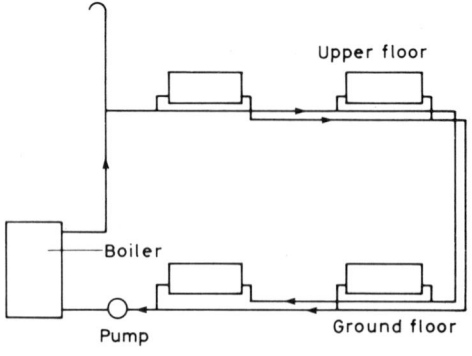

Figure 8.9. Typical single-circuit two-pipe system

shown in *Figure 8.9*. It is best to run in the way shown, taking the flow up and bringing it down as it moves on, rather than starting with the ground floor and then moving upstairs. In bungalows the up, over and down pattern is possible if pipe is run in the roof (*Figure 8.10*). But usually both flow and return are kept at or below ground-floor level.

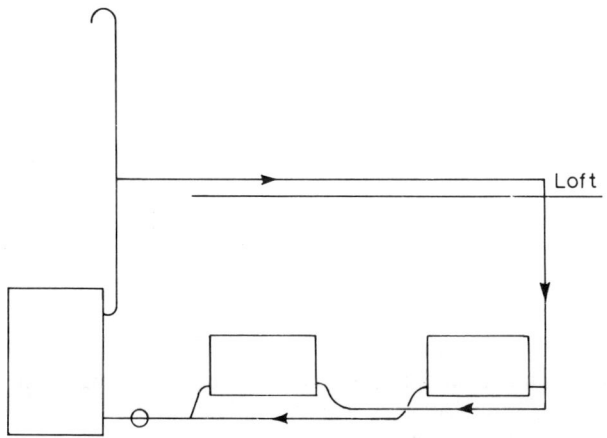

Figure 8.10. Typical 'up-and-over' pipe run for a bungalow

The second rough-design phase is to make a plan of the house, both or all floors, large enough for you to impose the principles given in *Figure 8.9* upon it. This is the first step in turning principle into practice. Do not be put off by the discovery that pipes seem to be going through solid walls, or past doorways. They will do this quite readily.

Next, survey the proposed route. Allow that most pipes will run under the floorboards, and if you can get runs with the joists it is very easy. There is no need to lift floorboards at this stage; just look at them. They run one way and joists run at right angles. It is at this stage that you have almost the last chance to reconsider. Perhaps a radiator could be moved, without breaking the basic rules of installation, but making pipe runs far easier? Then move it, now.

105

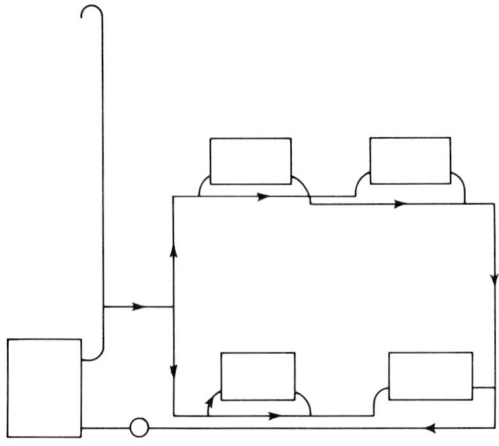

Figure 8.11. A typical split-level or zoned circuit which offers a chance to control automatically and independently the upstairs and downstairs (sleeping and living) rooms

We come at last to a final design, and it cannot be emphasised too strongly that this needs thought, and more thought. Visualise any snags, any possible improvements, because it will be expensive to change later.

The final design is the basis for sizing, remembering that skimping will lead to troubles through inability to pass enough heat, or, if pump-forced, by noise in pipes. Oversizing on the other hand is a way to waste money on pipe, and on heat loss from pipes.

Pipe-size calculation can be extremely mathematical or ridiculously simple. We hope to please most people by choosing the second way, since we could show in many ways that nothing about the entire process can justify a high degree of accuracy.

For each size of copper pipe there is a maximum figure, a quantity of hot water per unit time above which water noise can be expected. This volume can be expressed in the terms we already know, i.e. the heat demand per room, because the relationship between heat and water volume is based upon predictable working conditions, fairly steady boiler temperature etc.

The limits (to the nearest 0.5 kW) are, for

15 mm copper pipe,	5.5 kW
22 mm	11.5 kW
28 mm	21.0 kW

We will start by putting some figures to the radiators shown in *Figure 8.12*. This is a sketch from which inessentials in the present context have been omitted for simplicity. The rated emissions of the radiators or convectors shown are:

Radiator	1. 1.8 kW	4. 2.9 kW
	2. 2.7 kW	5. 1.8 kW
	3. 1.4 kW	6. 3.3 kW

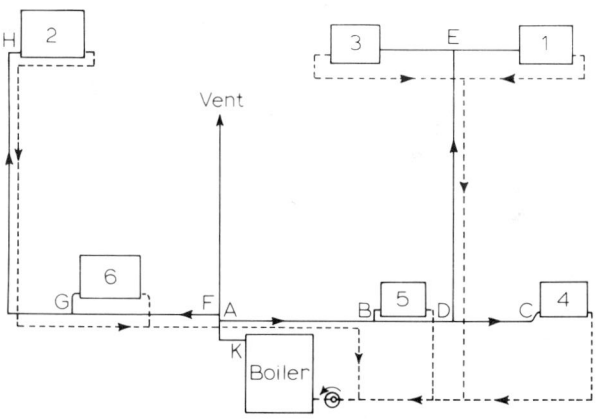

Figure 8.12. Layout of radiators

You will notice that the pipe circuits have been divided into sections according to the amount of work these do in supplying hot water (hence heat) to heat emitters on the downstream side. For example, pipe AB supplies all the heat for emitters 5, 4, 3 and 1. And so on. We can therefore put

107

figures to the duty performed by each section, and so, in terms of the rule quoted above, decide upon pipe size.

Section AB carries

1. 1.8 kW
3. 1.4 kW
4. 2.9 kW
5. 1.8 kW

———

7.9 kW.

7.9 kW is more than 5.5 but less than 11.5 kW. It is therefore beyond the capability of a 15 mm copper pipe but well within that of a 22 mm pipe.

Similarly, section BC carries

1. 1.8 kW
3. 1.4 kW
4. 2.9 kW

———

6.1 kW

This section also is in 22 mm copper pipe.

Section DE and its side branches carry

1. 1.0 kW
3. 1.4 kW

Total only 3.2 kW, requiring 15 mm copper pipe.

FG carries emitters 6 and 2, total 6.0 kW, needing a 22 mm pipe.

GH carries only emitter 2, 2.7 kW, another 15 mm pipe.

And last we come to FK, which feeds both AB and FG, total 13.9 kW. The first short length off the boiler is therefore 28 mm copper pipe.

Also in 28 mm pipe, but not shown on the diagram, are the flow and return pipes to the hot water cylinder.

The heating return-pipe sizes have not been calculated, since there is no need. If a certain volume of water flows to an apparatus, the same volume will flow away again, and the same size pipe will be needed. Thus, the return pipes will copy exactly the flow-pipe sizes.

108

Those pipe sizes can now be translated into lengths of each size in readiness for ordering. You are bound to allow a bit extra, because it is far more serious to find yourself short.

Pipe fittings are expensive, so try to make an accurate estimate of how many you need – you can always rush off and get another if you have to. For certain functions, bought fittings are essential: reducers (say from 22 mm to 15 mm); adaptors, from the iron thread of boilers and radiators to the compression joints of copper pipe; and tee pieces of any size. Note that at a place like D on the diagram, you can save one fitting by buying a reducing tee, to be specified as 15 off 22 × 22 mm copper compression. Or you can use a 15 off 15 × 22 mm tee and run section DC in 15 mm too. It shows the benefit of being adaptable.

Bends are another matter. Copper pipe in 15 and 22 mm sizes can be bent, thus forming slower bends and saving money. These pipes can also be kinked quite easily. So add a pipe bender to the list of tools you will hire. These include a copper-pipe cutter and a power drill with flexible drive, and masonry bit at least 25 mm diameter. You will need spanners for tightening fittings, a much smaller drill and bit for hanging radiator brackets. You could be wanting a blow lamp if you choose to have capillary fittings on the copper pipe. But although these are neat we do not advise their use, particularly if this is the first job. Compression fittings can be undone and remade. Problems about using a blowlamp in awkward, perhaps risky, situations are avoided. Screwdrivers and tools of that nature are usually to hand.

Buy an ample supply of pipe clips for all the sizes of pipe, for there are two important rules about pipe running. Make ample clearance through walls, timber etc. And fit a large number of pipe clips tightly, to prevent movement during expansion. An inadequate number of clips often leads to creaks and harsh noises during warming and cooling. But do not clip tightly near bends, for a bend can act as an expansion relief. For that reason, do not arrange for a bend to come up tight against a wall so that it cannot expand.

It does not matter whether you choose to start at the boiler and work on, or start at the radiators and work back. Measure

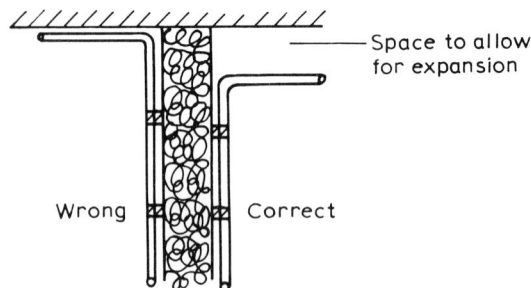

Space to allow
for expansion

Wrong   Correct

Figure 8.13. How to arrange pipes in corners

each length carefully, and allow for the amount of pipe which has to go into a fitting – capillary or compression – so that it will rest up against the internal shoulder. Another benefit of compression fittings is that you can make up several lengths of pipe, with several fittings, hand-tight only. This is much better than tightening as you go, only to find that the next length would go in more easily if the previous one were free to move. Save tightening for about two lengths back, when it seems certain that that part is complete.

Compression-fitting olives are a good fit, and a pipe with burr on it will not enter. So deburr pipe ends after using the cutter.

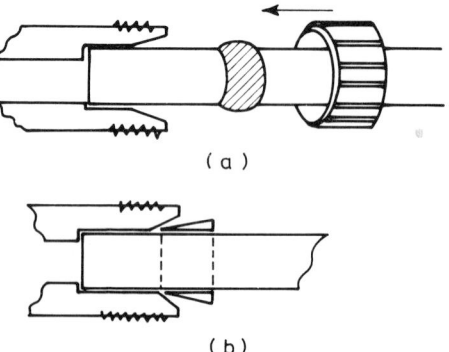

(a)

(b)

Figure 8.14. Compression joints with
(a) wedding ring, (b) wedge jointing

110

The method of making a compression joint is to fit the locking nut over the pipe end first. Follow this with the olive or similar compression piece. (Triangular ones have the chisel end pointing into the body of the fitting.) Enter the pipe into the body of the fitting, as far as it will go, up to the shoulder. Hold it there and push the olive into place. Follow with the nut and start tightening. Once you have bite, check that the pipe is still hard in place. When this is certain, tighten quite hard.

The principle of this kind of joint is that it squashes the soft-metal olive to give total close contact, and so the tightening has to deform the olive. This is a type of dry joint, but some people use a light smear of jointing compound over it before coupling and tightening, and there is nothing against it.

The necessity of being accurate about critical dimensions must be emphasised. For instance, the teed-off connection into a radiator must not be strained to fit, otherwise expansion will impose a great strain and could lead to leaking.

The pipework of a heating circuit is rarely insulated, and rarely needs it. Those pipes which are run within the house, for instance between floor and ceiling, make some contribution to house warming because their heat losses can go nowhere but in the house. Pipes which are run in the loft, or under the ground floor, are in a different category. If in the loft they should be covered by the loft insulation, not primarily to conserve heat but to prevent freezing should the heating be off. Pipes under the ground floor, on the other hand, are very prone to wasteful cooling in the draughty conditions which should prevail there. It is beneficial to insulate these pipes, either using pre-formed insulation or wrapping them in mineral-fibre blanket and binding with string. The same material should be used as a cushion or buffer around pipes which pass through the oversized holes we drilled during installation. A soft but inert lining of this kind prevents noisy movement during expansion, and keeps out insect life (*Figure 8.15*).

Every system has to be tested. Begin by applying a gravity test, simply allowing the system to fill with water from the

111

header tank, and examining each joint, from the bottom up. Once this proves satisfactory, try the pump, still on cold water. The pump adds a little to the internal pressure. A

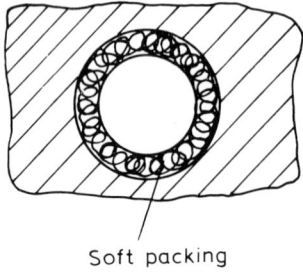

Soft packing

Figure 8.15. Holes through walls and joists

satisfactory outcome is hopeful, but the final test must come with high temperature added. For that reason you should not restore carpets and generally make good until the entire system is effectively at work.

Microbore

The principal feature of microbore, the very small bore tubing, brings with it a special difficulty. The job of heating requires the same amount of heat, whatever the method, so that the same quantity of heated water must be passed regardless of tube diameter. Resistance to water flow increases sharply as diameter decreases, and for a start this calls for a more powerful pump. It then requires that pressure absorption shall not be cumulative, item to item, which means that series flow, such as occurs with a one-pipe small-bore system, cannot be used. It has to be parallel flow, which is a feature of a two-pipe small-bore system. But with microbore it is more usual to arrange it as in *Figure 8.16*, each emitter being served by an out-and-home pipe from a pair of manifolds respectively flow and return.

The similarity of this method to the wiring of electric storage heaters is heightened by the fact that microbore tubing can itself be handled very much like heavy cable.

Microbore tube for the average house comes in three sizes, 8 mm, 10 mm and 12 mm, which roughly correspond to the 15, 22 and 28 mm of small bore. But one of the makers of tube, the Wednesbury Tube Co., has printed a booklet which goes into the details, and anyone seriously considering microbore ought to get a copy.

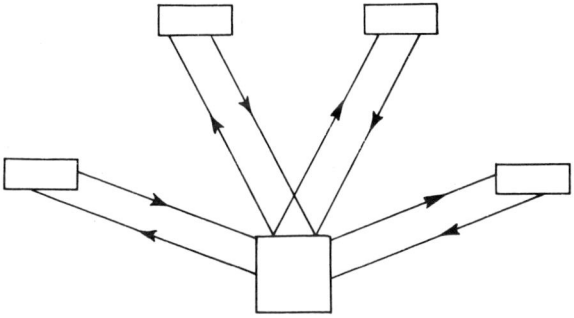

Figure 8.16. The principle of microbore

The radiators are chosen and located just as for small bore, and indeed functionally there is no difference. But the pipework is quite different. It might help if you look upon it as electric cable, and then to imagine that each radiator has to be wired back separately to a common fuse box. Then you have to decide where the 'fuse box' should be situated so that it is readily accessible to all the radiators, not a very long way from some and very close to others. Once a suitable place has been chosen, remember that it is not a fuse box but a pair of manifolds for flow and return. And the next question is, how accessible is the place chosen in relation to the boiler? If it is a very long way away, for instance at the other end of the house, the best policy is to seek another meeting place for the microbore pipes, nearer to the boiler. The general proportions desired are shown in *Figure 8.17*, with the added complication of being zoned for upstairs and downstairs. The manifolds A and B are connected to the boiler by pipe of small-bore size, in the illustration 22 mm. Also shown are the twin-entry valves commonly supplied for this purpose, for neatness. They have twin waterways in a common body.

It is usual to limit microbore runs to 9 m (30 ft), so that in larger houses there might be more than one manifold per floor. The manifolds should be located in pairs so that each flow and return pipe in a pair of microbores is of about the same length.

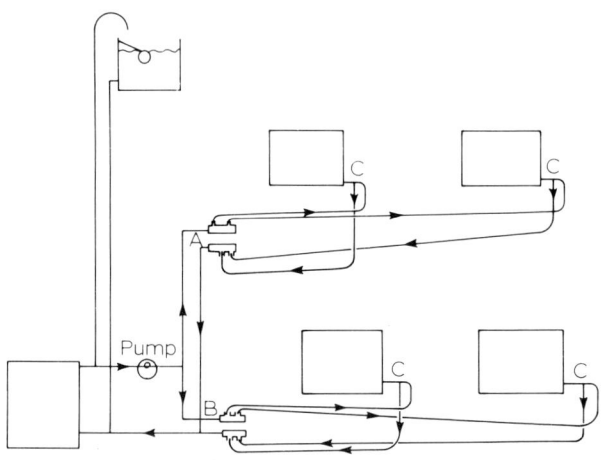

Figure 8.17. General layout of microbore system. (A) Upper-floor manifolds. (B) Lower-floor manifolds. (C) Combined inlet/outlet radiator connections

It would be wrong to look upon microbore as heralding the new technology to the exclusion of small-bore. Each has a place. The advantages of microbore may be summed up as:
• rapid warm-up because of the small amount of water in circuit;
• ease of installation of the pipe; and
• possible economies in material and installation cost. (But do not be persuaded to use nylon pipe, however cheap.)

Likely drawbacks are greater ease of blockage by large particles, and the need to use a more powerful pump.

Domestic hot water

Little mention has so far been made of the domestic hot water side, though when it is included it is wholly integrated, and interdependent with other features. Among the more important items to note are:

● The hot water cylinder should be a fully indirect type. A self-priming model is permissible as second best, but NEVER use a direct cylinder.

● An indirect cylinder calls for two cisterns in the loft. There is the feed/storage cistern to supply water to the hot taps via the cylinder. And there is the header tank or feed/expansion cistern which takes care of the primaries, water in boiler etc. The first is usually 25 gallons or just over 100 litres, the second, minimum 20 litres (4 gallons) is up to 45 litres (10 gallons). Both should be mounted as high as possible, even on a trestle under the apex if the general configuration of the dwelling is low. The first needs height in order to promote a good flow at taps and to create a satisfactory shower bath. The second should be removed vertically as far as possible from the influence of the circulating pump to 'pump over' – see Chapter 9. (That is an argument against 'up-and-over' heating pipes running in the loft.)

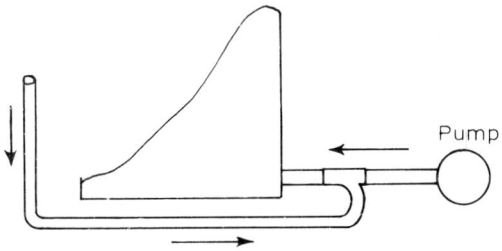

Figure 8.18. Common return to the boiler

● The 28 mm (1 in) flow pipe may be taken from either boiler flow tapping, usually not the same one as the heating flow uses. Unless the boiler chosen is specifically designed, with an internal partition, to prevent cross influence between

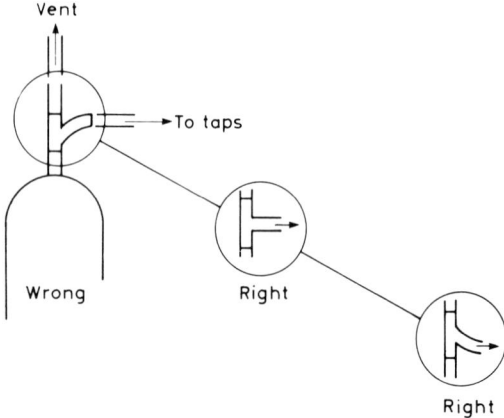

Figure 8.19. The right and wrong ways of fitting the tee for hot water supply to the cylinder outlet

opposite return ports, the primary return should be brought in together with the heating return. See *Figure 8.18*. Note the use of a swept tee and the position of the pump.

• The offtake from the riser off the cylinder, a tee piece, must be either a plain tee or a reversed swept tee. It must not be a conventional swept tee, which would encourage air locking. See *Figure 8.19*.

• The hot water cylinder must be insulated to a very high standard.

9

Power units for wet systems

The blackleaded cast-iron object in the kitchen is no more, and what have succeeded it are not always recognisable as boilers (boiler being the wholly misleading name given to the unit which heats the water in hot water systems). One way of classifying the present generation of boilers is by location and consequently the amount of useful space they occupy, since that is sometimes important. The most unobtrusive to the point of being invisible is the back boiler, which takes no useful space at all, being placed inside the hearth of an ordinary fireplace.

Next in order of unobtrusiveness comes the boiler which occupies no floor space, being mounted on a wall and looking like a familiar multipoint water heater or a hanging cupboard. And last is the floor-standing boiler, for which strenuous efforts are sometimes made to lessen its floor space, by slimming down and/or shortening it so that it will fit under a draining board, and so on.

Another criterion for gas might be type of flue, whether open (conventional) or balanced, though it would be misleading to think of this as an area for deep deliberation. There are no really significant differences, and as much as anything it should hinge upon availability or convenience. If there is no chimney, or if the preferred position for a boiler is not by the chimney, then the balanced flue through an outside wall is possible, except for fire back boilers.

Potential heat output is another source of choice. Firebacks have a physical limit of about 13 kW (45 000 Btu/h)

117

because of restricted space. Higher outputs may, however, be expected in due course. Wall-mounted boilers go higher than 22 kW, but floor-standing units go to the maximum nominal domestic limit of 44 kW (150 000 Btu/h).

Choice of fuel introduces an element of variety even within the choice offered by the three forms. All three forms, solid fuel, gas and oil, offer a back-boiler arrangement. But oil has only a very limited range of units to offer. Solid-fuel back boilers can do anything from domestic hot water only, to a fair amount of central heating, up to about 8.8 kW (30 000 Btu/h) from selected models. Gas goes up to 12 kW (40 000 Btu/h) and in at least one case to 13 kW (45 000 Btu/h). Another difference lies in mode of operation. Gas and oil can run the boiler independently of the fire in the front. Solid fuel uses the fire in the front, making an auxiliary system for summer a worthwhile consideration.

Wall-mounted boilers are always gas boilers, and in this species we find an interesting division, which could constitute a valid area for choice. On the one hand we have makers who have taken their successful floor-standing boilers, usually with cast-iron heat exchanger, and all the expected control system, and have simply hung it on a back-plate for wall attachment. Makers such as Baxi argue that reliability with an adequate performance is what matters most, and those virtues have been well proven by their standard unit. Others, among them Thorn EMI, use the wall-hung boiler in order to employ the latest technology, of lightweight heat exchangers and low water content giving rapid warm-up and much reduced losses when the boiler is cycling. The price to pay for this is a certain technical touchiness. Should anything go wrong with the water circulation at high gas rate and low water content, trouble would follow quickly. Consequently there are some precautionary techniques: fully pumped system and high head pump, system bypass, controls which start the pump before gas is lit, and so on. This type of boiler is really out of the class which can be put in by an amateur, though of course when competently installed its benefits are obtained.

Continuing the search for comparisons, it used to be

118

possible to choose between those depending upon electricity and those which do not. But nowadays the field has narrowed. The only boilers which almost certainly do not have any electrical connection are solid-fuel back boilers. And as we shall see when discussing controls, once you introduce electricity there is no limit to what you can do, easily and moderately cheaply.

These are the main areas of comparison which are worth studying in order to see not only what is available but also where it would be a waste of time to look. There are certain conditions which do the choosing for you. Here are a few examples.

The remote country cottage with no power, gas or electric, can have nothing but solid fuel (except that bottled gas could run a non-electric fire). If water is bucket-drawn, no wet system is possible, whatever power is there.

The same cottage with electricity can have storage radiators and (given a water system) an immersion heater.

A flat cannot house a conventional flue appliance nor, generally, a system which depends upon having a water header tank at a good working height.

There are still places, not too remote, which have no mains gas. It is possible to have a full gas heating system using, at least temporarily, bottled gas burned in appropriately adapted appliances.

There is a small number of people devoted to wood burning because they get the wood free, but this is not strictly a technical reason.

The foregoing list of comparative features brings out several points which need to be kept in mind by anyone who is thinking about a new boiler. It might be thought that certain of them are quite positive, for instance that no back-boiler can possibly supply a full heating system for a six-bedroom house. But that is true only at present. One cannot say what will be achieved in development laboratories, and when a need arises it is worth asking.

Perhaps the most significant point to emerge is the amount of interrelation. It does not matter, technically, to the principal purpose of having a heating system, whether you choose oil, gas or solid fuel, wall-mounted or floor-standing, balanced or open flue, high or conventional technology. These and other matters are more bound up with personal preference and local issues like service availability and site limitations.

Installation

When we come to installation procedures we find that all modern boilers have at least one thing in common. All must be connected on the hot water side to *an indirect cylinder*. Some will accept a semi-indirect or self-priming cylinder, but in these cases it must be chosen and installed strictly as recommended by the maker. There is one lamentable exception. Certain solid-fuel back-boilers do not have to have an indirect cylinder, but we must urge readers to behave as though they do.

Site requirements
Back boilers. A suitable fireplace will have not less than standard dimensions, i.e. width 406 mm (16 in), depth 343 mm (13½ in), height 543 mm (21⅜ in).

These are basic dimensions, the fireplace after cheeks etc. have been removed. There are maxima given as well, but fireplaces can be made smaller if necessary. The chimney must produce a constant adequate draught.

Wall-mounted boilers. Two things matter about the wall, the situation and the construction. It must be an outside wall (for a balanced flue) or within easy access of an open flue, new or existing.

As for construction, the most important factor is strength. Makers supply back plates or brackets which are intrinsically adequate, but whether they will stay on the wall is entirely up to the installer. Outside walls generally are quite strong enough. Wall plugs must be drilled back into solid brick or stone, not left in a plaster face.

A wall which is classed as combustible, e.g. timber, is not barred. But certain fireproofing measures must be taken, e.g.

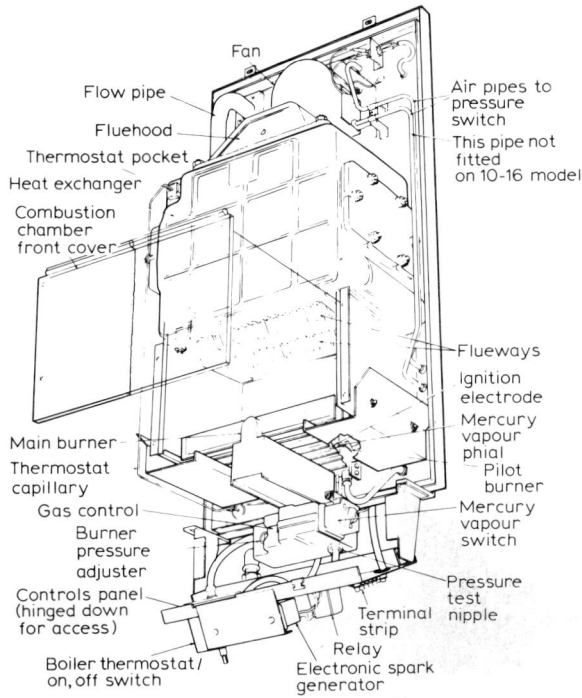

Fan

Flow pipe

Fluehood

Thermostat pocket

Heat exchanger

Combustion
chamber
front cover

Main burner

Thermostat
capillary

Gas control

Burner
pressure
adjuster

Controls panel
(hinged down
for access)

Boiler thermostat/
on, off switch

Air pipes to
pressure
switch

This pipe not
fitted
on 10-16 model

Flueways

Ignition
electrode

Mercury
vapour
phial

Pilot
burner

Mercury
vapour
switch

Pressure
test
nipple

Terminal
strip

Relay

Electronic spark
generator

Figure 9.1. General arrangement of the Potterton Neatheat
wall-mounted gas-fired boiler. It is a balanced-flue boiler, and
the combustion chamber front cover has been removed to
show the flueways

inserting a slab of insulation behind the back plate. If you
have such a wall, it is desirable to obtain the advice of the
local authority building department, or at least of the manufacturer. Wall-mounted boilers must never hang on their
pipework.

Floor-standing boilers. As with wall-mounted boilers, an
outside wall or access to an open flue are needed, also
permanent space for maintenance to limits which the maker
specifies, e.g. for removal of the heat exchanger. The required floor condition varies. It must, of course, be strong

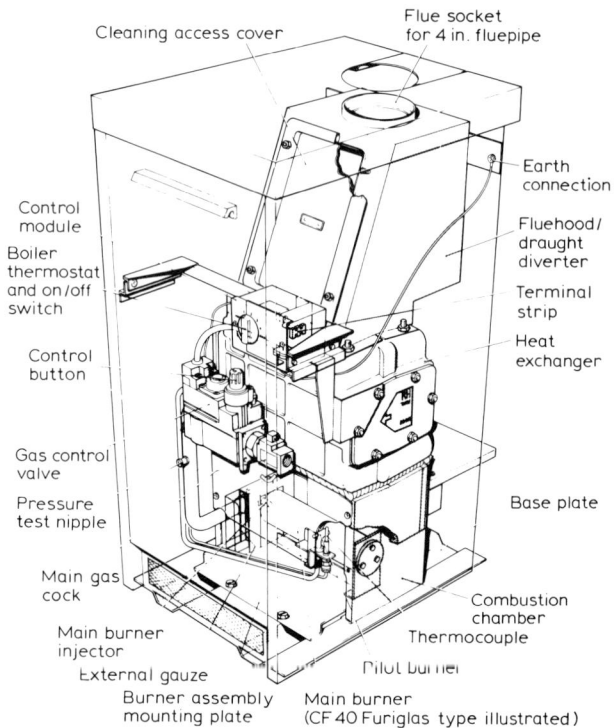

Cleaning access cover

Flue socket
for 4 in. fluepipe

Control
module

Boiler
thermostat
and on/off
switch

Control
button

Gas control
valve

Pressure
test nipple

Main gas
cock

Main burner
injector

External gauze

Burner assembly
mounting plate

Earth
connection

Fluehood/
draught
diverter

Terminal
strip

Heat
exchanger

Base plate

Combustion
chamber

Thermocouple

Pilot burner

Main burner
(CF 40 Furiglas type illustrated)

Figure 9.2. General arrangement of the Potterton Kingfisher floor-standing gas-fired boiler for open flue. There are balanced-flue versions of the same range of boilers

enough for the usually quite-considerable weight. But only the boiler can determine whether it must be fireproof.

It may be taken that all solid-fuel boilers must have a firm and incombustible base such as concrete or brick would provide. Similarly with some gas- and oil-fired boilers. But if a boiler is so constructed as to have a wrap-around boiler, with a waterway at the base, then it follows that the base temperature can at no time exceed that of boiling water, 100°C, which is not dangerous. The most it can do is to soften plastic and

122

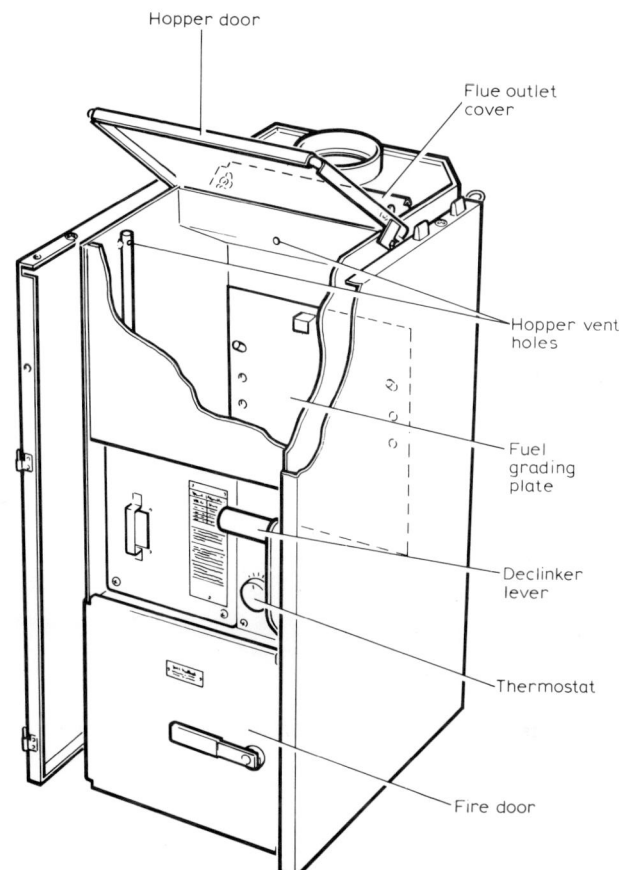

Figure 9.3. General arrangement of Thorn EMI Janitor solid-fuel boiler

lino tiles or sheet, and these should be protected by standing the boiler on a base plate, conveniently made of a steel sheet on an insulator such as mineral-fibre blanket.

Ventilation
The whole subject has been covered in some detail in Chapter 5, and again in Chapter 6. The fundamental rules set

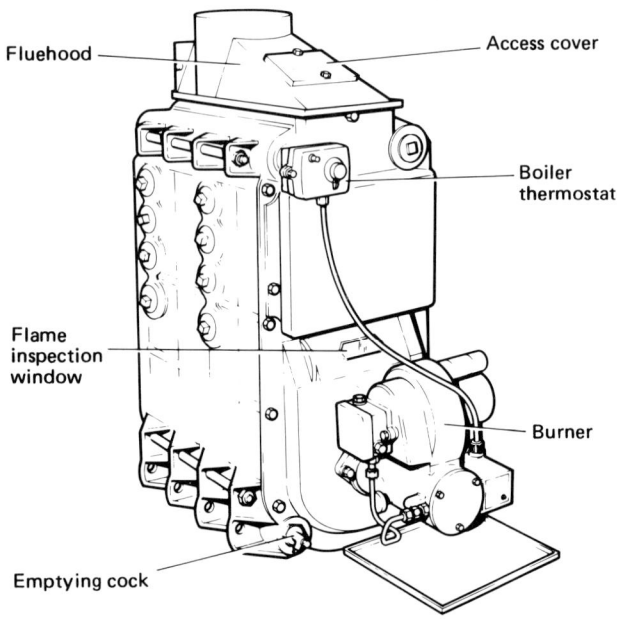

Fluehood

Access cover

Boiler
thermostat

Flame
inspection
window

Burner

Emptying cock

Figure 9.4. Potterton BOA Series 4 pressure-jet oil-fired boiler
shown with the case removed

out there apply to all the types of boiler, until we reach the
special cases of fanned-flue boilers, which are beyond the
scope of this book. Similarly, Chapter 5 deals with flues of
both open and balanced kinds.

Water connections
For back boilers, adequate holes must be made through the
side wall or walls of the chimney breast *before* the boiler is
put into position. The generous allowance, which can be
patched up later, is far better than trying to make accurate
assessments of position. A drain cock is usually fitted into the
lowest pipe leaving the boiler, on the outside of the breast
wall. At the early stage only first lengths need to be fitted. But
the entire system should be made up, and tested, before
steps are taken to fit any of the fronting apparatus.

124

For other boilers, cased versions must first have the case removed. Wall-mounted boilers are fitted on to their back plate, floor-standing boilers are set firm and level on the base. In both cases the primary need in location is to marry easily to the flue, of whatever type. When that is achieved the water connections can be made. If local authority regulations require a safety valve, this should be fitted in the flow pipe as near to the boiler as possible.

Gas connection
A gas supply pipe of adequate diameter to supply the boiler as well as any other appliances downstream of the boiler must be used, and connected via a standard stopcock. Test as described in Code of Practice 331:3.

Electrical connection
In most cases a single phase 3-wire supply a.c. of 50 Hz, with fused double-pole switch or unswitched fused plug and

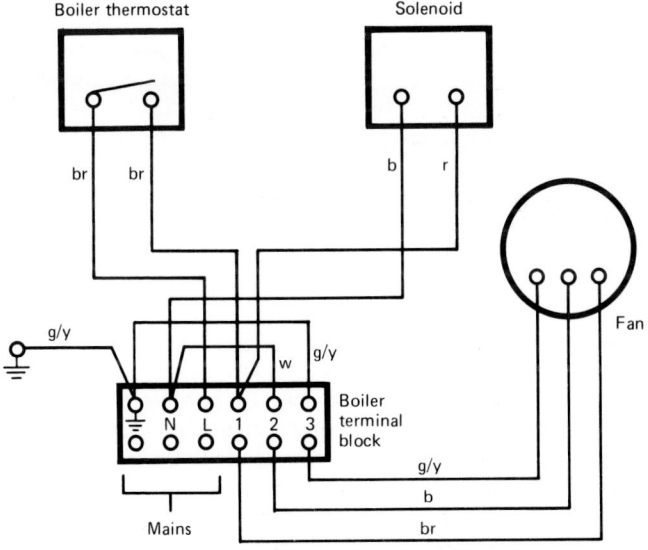

Figure 9.5. Simple wiring diagram for the Thorn EMI Janitor solid-fuel boiler

socket, is called for. A fuse rating of 3 amp is sufficient for almost all boilers, but it might have to be increased to include a circulating pump. See *Figure 9.5.*

Other items
In some boilers the instrument panel or its equivalent, a control box, is left off for convenience of the rest of the installing. It should be fitted next, if attached to the boiler. Or, if it is to be case-mounted, the case must come next. In either event it is rare for any complex wiring to be needed since all connections are included within a composite plug/socket fitting. Similarly, external fittings and controls are usually accommodated by having plug-in points left for them on a control board. If any one thus allowed for is not fitted then the terminal points must be bridged.

The above notes have tried to give a general view of what is to be expected in installing a boiler of any type, and are only a preparation, not a substitute, for reading the precise instructions issued by the manufacturer and supplied with his product. There is no need to fear that manufacturers' literature might be inadequate. We are considering throughout this book boilers which have been officially tested and approved by the appropriate testing authority, and the instructional literature is in all cases regarded as an integral part of the appliance. There have been cases in which boilers have failed to gain approval until the literature was brought up to standard.

Commissioning equipment

A law suit, famous in its day, was begun by a man who followed the boiler commissioning instructions, which read 'fill the system, run the pump, then drain to flush out foreign matter'. The instructions assumed, wrongly, that the man would then refill the system. He did not, and the boiler was burned out. His action contended that he followed the instructions, and he won the case. But the moral in that tale is for the manufacturer, not for the purchaser and reader. The event serves to show that commissioning is very important. It depends first upon satisfactory installation. It then requires, in a condensed timescale, knowledge and experience which in the case of the more technical products it would be unrealistic to look for in the amateur. And, as we have pointed out elsewhere, there could be a lot of capital and an alarming amount of domestic safety at stake.

We will look in outline at what is required in the commissioning of the various types of equipment we have been discussing. Let us begin with the premise that in every case the system has been filled, tested for leaks, drained and flushed, and refilled, with some attention to eliminating air pockets.

Solid-fuel back boiler

This requires no special attention. The correct grade of fuel must be stored and used on the fire, and the flueway from the fire to the boiler must be clean of all debris, and the damper working. Starting from cold, with a cold chimney, the damper should be in the 'fire only' position to start, in order to give the fire the full benefit of chimney pull. Only when the fire is established may the damper be moved to 'boiler' position.

Gas-fired back boiler

Once the boiler is satisfactorily installed, send for the gas man, with his special skills and equipment.

The spatial relationship between pilot burner, thermocouple probe and igniter electrode is fixed and precise and must never be interfered with, on any gas-fired appliance.

At this point check the burner pressure, first taking the cap off the pressure test point, then fitting on a pressure gauge. The correct pressure is given in the instructions and must be used. There is a fixed relationship between injector size and burner pressure on all gas-fired appliances. It determines the rate of heat input which has been decided in order to get maximum performance. Refit the cap on the pressure test point.

Relight the pilot burner, switch on or plug in electricity, and turn the boiler thermostat (and clock if fitted) to On. The boiler is now in working order, and the fire may be fitted.

Gas-fired wall-mounted boiler
The simpler version of this apparatus, the boiler which is a rehoused version of a floor-standing boiler, will be put to work in roughly the same way as the back boiler. Consistent with our advice that the more complex gas-fired boiler, with low water capacity, demands expert attention, we will not

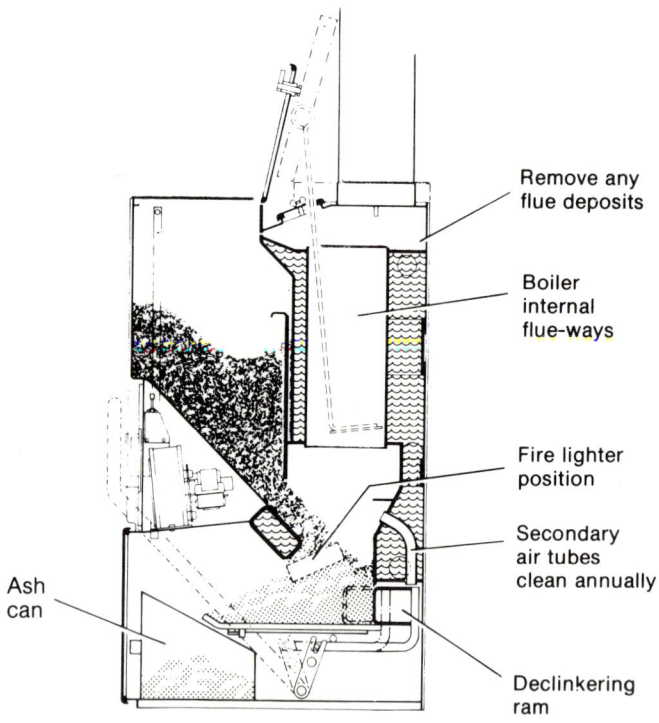

Figure 9.6. Schematic detail of Thorn EMI Janitor solid-fuel floor-standing boiler. Note: the hopper lid, shown open, would be tightly closed during normal running

attempt to précis the commissioning procedure. But it is detailed in the literature supplied with each boiler.

Solid-fuel floor-standing boiler

The type we show is a hopper-fed unit with fan-powered draught and thermostat control. It is possible to get larger and more complex units, for example having screw feed for fuel. For initial lighting, leave the hopper empty. Take off the fire door, and start a small fire, with paper, wood and fuel or gas poker and fuel. If necessary start the flue with a piece of burning paper. When the fire is well established, close the fire door and introduce fuel via the hopper. The hopper lid or door must be kept tightly closed and its lute free from foreign matter.

Switch on electricity and control the boiler by means of the thermostat. As warming proceeds, trace its progress along the flow pipe to the hot water cylinder, and make sure in due course that warmth is being given off by the permanent standing loss, which is probably a radiator or towel rail in the bathroom. This is the thermal safety outlet and must *never* be closed off.

Oil-fired floor-standing boiler

Begin by purging the oil pipeline of water and foreign matter. Break into the line twice, once at the filter inlet – run off a bucketful and reconnect – and then do the same at the boiler. After settlement the buckets of oil may be returned to the tank. Make sure that any controls, thermostat and clock, are set to On. Then, for a pressure-jet burner, press the reset button in case it has tripped, and switch on the electricity. (For vaporising burners the appropriate operation is described in the literature.) Being energised, the control box will attempt to light the boiler. If a stable flame is not established within about 30 seconds, control will go to lockout. Press the reset to try again. Continued refusal indicates a fault such as fuel shortage or a dirty photocell window.

After 15 minutes steady running you must conduct some important tests. Pump pressure, flue gas temperature, CO_2

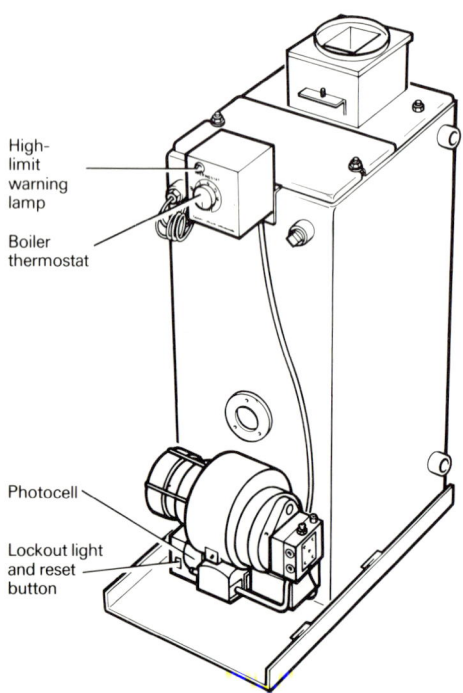

High-limit warning lamp

Boiler thermostat

Photocell

Lockout light and reset button

Figure 9.7. The Thorn EMI Panda oil-fired pressure-jet boiler shown without case

and smoke number must be found and checked against the datum figures supplied. These figures are crucial to successful running. For instance, if smoke is higher than allowed, it will quickly lead to a smoke-obscured photocell, and automatic shut-down, and before long will result in a sooted combustion chamber. Setting up an oil burner, as it is called, is a job for professionals with appropriate equipment.

Gas-fired floor-standing boiler
The need for professional help with commissioning is as great as with the gas-fired back boiler. The floor-standing

boiler has some advantage in being more accessible. Perhaps we might mention a feature of routine lighting which is not always made clear. All boilers include what is called a flame-failure device, by means of which, if for any reason at all the flame, usually the pilot, were to go out, the entire gas supply to the burner would be cut off. Lighting a burner from cold would therefore be impossible but for a special feature. It make take more than one form, but commonly it is a knob on the multi-functional gas-control valve. Let us suppose that this knob is marked, and can be turned to Off, Pilot and On, and turning to and from On must be depressed slightly, to avoid inadvertent turning off. Now, to light the pilot flame, turn this knob to Pilot. Then push it hard in, and operate the piezo lighter knob (or apply a lighted taper) until the gas lights. Continue to hold the knob hard in for up to 30 seconds, then release slowly, and the pilot should stay alight. Then turn the knob to On. And the reason for holding down that knob is that it overrides the automatic response of the control to shut off the gas, until the thermocouple has warmed up and can take over the duty.

Maintenance

Maintenance is as necessary for these high-efficiency machines as it is for your car, and for much the same reasons. We will deal with the apparatus in the same order as before.

Solid-fuel back boiler
The only attention needed is to flue ways. These should be thoroughly swept at regular intervals, and the chimney itself cleaned at, usually, three- to four-month intervals or as seems necessary. One should not neglect care of the fire itself, and in particular of the ceramic cheeks and back plate. If these crack they should be repaired using either fireclay or fire cement, the latter often sold in a plastic condition ready to apply.

Gas-fired back boiler

There are two maintenance jobs here, one on the boiler and one on the fire. If we deal with the boiler it will stand also for the conventional pattern wall-mounted and the floor-standing boilers.

Time and use bring two kinds of nuisance to gas-fired boilers. One is iron scale on the outside, i.e. flue-gas side of the heat exchanger, and this is removed by brisk brushing, having first taken off the top plate and the burner assembly. The other is called lint – dust and fibres airborne to the burner along with combustion air. It collects and can block or partially block airways, and drastically alter combustion characteristics. So burners have to be thoroughly cleaned. But – to repeat – preserve the pilot/thermocouple/igniter relationship unchanged.

Oil-fired boiler

Cleaning the oil filter is a regular item, not one which should wait for maintenance. This job involves considerable stripping, so that the combustion chamber and any baffles in it

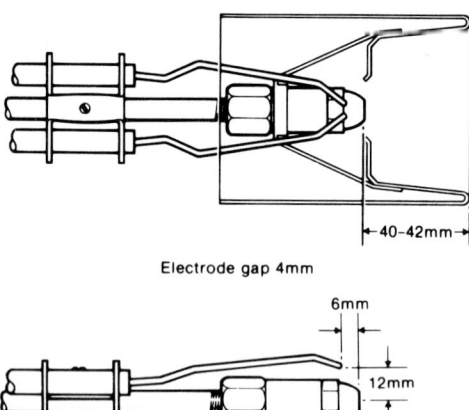

Electrode gap 4mm

Figure 9.8. Typical electrode settings for a pressure-jet oil-fired boiler. These relate to the Thorn Panda boiler

may be thoroughly scraped and cleaned, then replaced in correct order. The photocell and its equipment has to be cleaned. Another important task is to check and if necessary adjust the electrode gaps in accordance with makers' instructions. A typical setting is shown in *Figure 9.8*.

It is usual to split and clean the fan, which collects airborne dust, and the pump, where oil dirt could lodge. When it has all been reassembled, the same checks must be applied as were detailed for commissioning.

10

Heating by dry systems

All central heating is concerned with the warming of air. The wet systems go about it indirectly, by warming objects like radiators which then warm the air. Despite the many virtues of wet systems, it takes a fairly long time to achieve the initial warm-up and to bring about any changes of air temperature in accordance with instructions from the user or the control system.

Dry systems, by contrast, are direct. You want warm air, they supply warm air. If you stand in the path of the discharge it is almost instantaneous warming, just as changes are noticed very quickly. Even if you are not in the direct air stream, room temperature rises or changes quite quickly, simply because the heater is supplying the room with a complete consignment of ready-warmed air. So, bearing in mind that most heaters can also produce domestic hot water, and the electric ones can share their off-peak rate with domestic hot water, why does anyone bother with wet systems?

A fundamental objection to dry systems lies in their having not a scrap of radiant heat, and at the end of this chapter there is a note in explanation of the importance of radiation. Wet systems with radiators do produce quite a lot of radiation.

But there is a practical objection too, which effectively bars their general adoption in existing houses. A 15 mm (½ in) pipe can carry as much heat in water as an air duct measuring several inches square. And there are very few houses which could tolerate air ducts of that size running at floor or ceiling height, as well as a return duct even bigger. So, generally speaking, full-scale ducted systems are restricted to new premises where the architect and the builder can contrive to 'lose' the ducts in walls, under floors, and generally in the structure.

In a book of this kind we need only remind any reader lucky enough to be having a house built that the time to start asking for ducted heating is at drawing-board stage. After that it might be too late. Suppose, though, that one were called upon to devise a heating scheme for the local scouts' hut or similar open-plan building where some small resemblance to the engine room of a submarine would not be

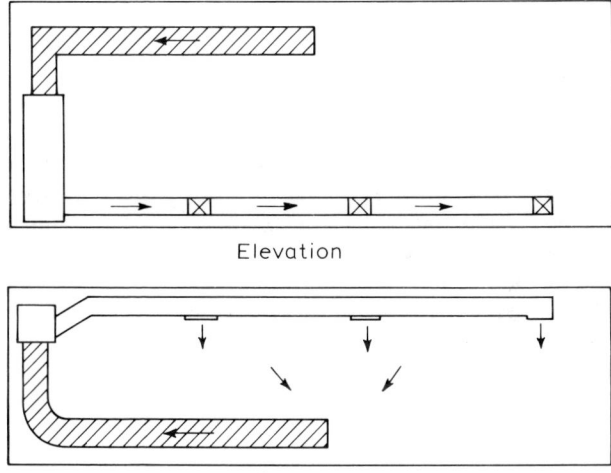

Elevation

Plan

Figure 10.1. Main features of a ducted warm-air system

amiss. Here, a system which could provide rapid warming from cold on meeting nights would be best. So a ducted warm-air system such as the one shown in *Figure 10.1*, would be very suitable.

There are, for a start, varying ideas about where warm-air discharge should take place on the vertical scale. These range literally from floor to ceiling: floor or 'perimeter' outlets or registers (duct outlets fitted with adjustable dampers) which discharge upwards; ceiling registers which project downwards; low-level wall registers such as *Figure 10.1* shows. The first give good results but tend to be rather vulnerable and act as dirt traps. The second are more popular in North America and tend to create hot high levels, which are not the real object. That is why we show the low-wall type, a very useful compromise.

As for the duct itself, the installation under review has no need of calculation. We need to use the maximum of warm air, and there will be no more of this than can be carried by the heater outlet port. So we will make our duct to that size. (It could be stepped down after intermediate offtake points.) We show three registers spaced along the room, which we assume to be some 10 to 12 m long (30–40 ft). The registers enable the user to shut off chosen outlets, and also, if no intermediate damper is fitted, to so control emission from the early outlets as to make ample air available towards the end of the duct. The system must be balanced so that all outlet points can get their fair share of the total warm air.

Next you will notice the high-level return air duct. In some cases this cannot be left out, notably where the air heater is gas-fired to an open flue. In all cases it should not be left out, for that would interfere with a satisfactory pattern of air movement. In *Figure 10.1* the single pick-up point for return air is centrally placed in the building, which makes for uniform air movement.

We could go on to extrapolate from *Figure 10.1* enough detail for our simple system to be suitable for a bungalow. Let us suppose that instead of having three registers let into the warm air duct, we fit branch ducts and transfer the registers to the ends of these ducts. Not much has altered on the system, but the discharge points have been taken to other parts of the building. We could even divide one or more of these branch ducts, with a tee, to give a pair, with registers. The branch ducts, carrying only a portion of the total warm air, would be of smaller cross-sectional area. And that, in brief, is how a system is created.

136

Other points to note are:

- The air ducts must be insulated. A very suitable insulator is resin-bonded glass fibre, mitred to fit and held on by tape.
- The return air duct is not insulated.
- Heaters may be downflow, upflow or crossflow units. These terms relate to the direction of air travel within the unit. The one shown in *Figure 10.1* is a downflow unit, the fan being at the top.
- In a compartmentalised building, i.e. having rooms with closeable doors, it is not necessary to extend the return air duct to every room. Given a small number of central pick-up points, air may gain access via relief openings cut from the tops of doors.
- Return air must not be taken from kitchen, w.c. or bathroom.
- A filter is fitted in the return-air spigot on the unit and possible additions are a humidifier, a secondary electrostatic filter, and a means of admitting fresh outdoor air.
- The unit may be used to circulate cooling air in summer.
- Gas fired units may incorporate a water heater, in which case this will be independently operated but sharing a common gas supply and flue.
- The gas industry has issued a Code of Practice concerning warm air installations, and this is worth studying.

Indirect heating

So far in this chapter we have been concerned with what is called direct heating. This is a comparative term, since true direct heating, such as is used on building sites, dispenses with a heat exchanger. Hot air and hot flue gases come out together. This method is not suitable for domestic premises. In this context direct heating means that heat from the fuel heats the air through the medium of a heat exchanger. Its counterpart, indirect heating, starts not with an air heater but with a boiler. Hot water flows from the boiler to a special heat exchanger which is in effect the inside of an air heater, except that it is hot water and not flue gases which flow through the heat exchanger. The fan and air flow are as for the direct unit.

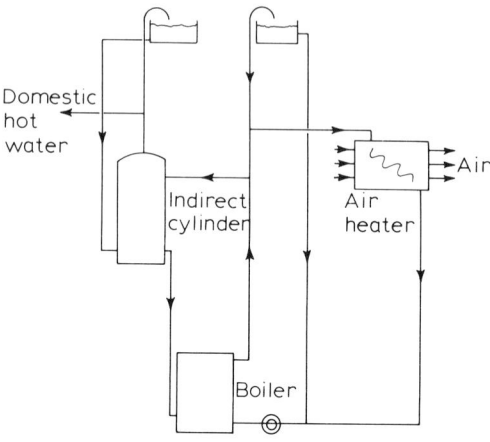

Figure 10.2. System with indirect air heater

The circumstances in which an indirect unit might be used are, for instance, where the most convenient place for the air heater is wholly unsuited to flue or fuel access, or to a combustion operation. The solution is then to transfer the 'engine room' to a suitable place, leaving the heat exchanger in its best position.

Another possibility, for which more than one manufacturer caters, is that an existing air heater is worn out but all its duct connections are too good to disturb. The manufacturer will then provide a heat-exchanger unit with facsimile connections, which will be served with energy from a boiler in another location. There is indeed a choice of reasons why an indirect system might be worth considering, whether for a full system or a partial or selective one.

Figure 10.3 shows a typical downflow air heater complete with optional water heater, while *Figure 10.4* shows an indirect heater unit.

138

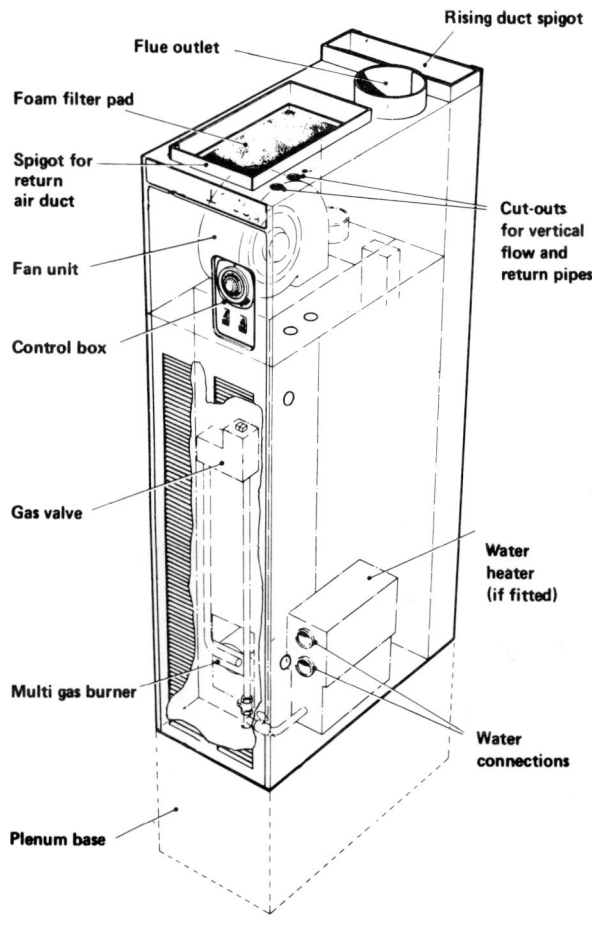

Figure 10.3. An example of a downflow air heater with optional water heater (courtesy Thorn Heating Ltd)

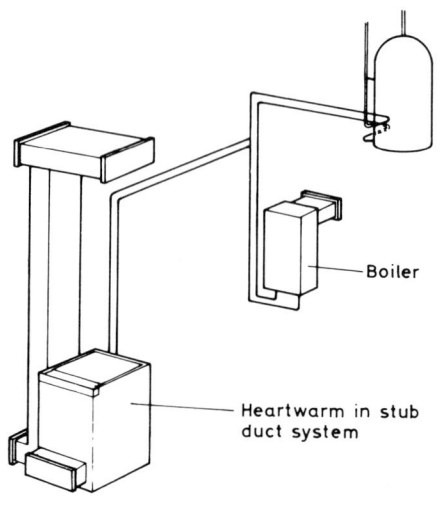

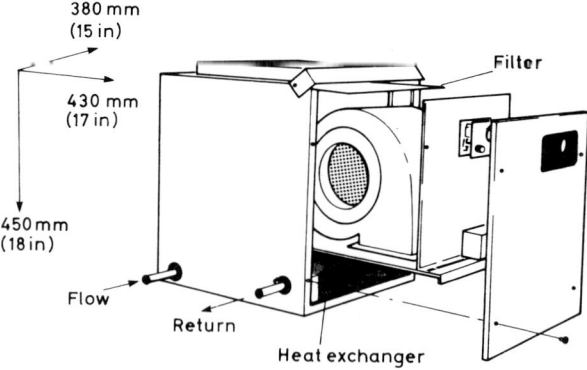

Figure 10.4. The Heartwarm air/water heater for use in an indirect system, showing the hot water flow and return connections (courtesy Massrealm Ltd)

Stub duct system

If we have some practical involvement it will be with a stub duct system. This is not different from a full duct system except in scale. Being reduced in scale means that the duct sizes, though large, are less than for a full system. But principally it means that a special feature of the choice of rooms to be warmed is that they are, either horizontally or vertically or both, adjacent. In this way short and mainly out-of-sight routes can be devised for the ducts, and this is shown in *Figure 10.5*. There are obviously more variations upon this theme than there are different house plans, and each house must be examined to see what can be made of it.

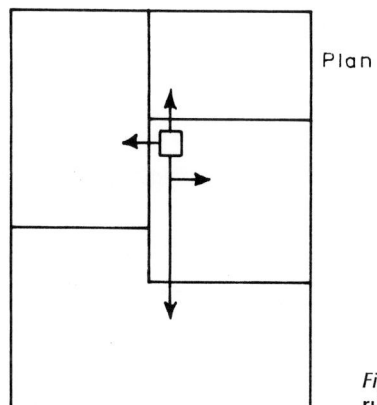

Plan

Figure 10.5. Typical stub duct run

For the non-professional it is perhaps fortuitous that detailed calculations of duct size are not called for in stub duct systems. One of the resemblances they share with the scheme shown in *Figure 10.1* is that ducts of outlet spigot size will be suitable.

The rules for ductwork are as before. All ducts must be made airtight, which means using jointing and taping joints, before applying insulation. Each terminal end must be fitted

141

with a register, and it pays to choose registers with a ready means for closing and opening. The more a system of this kind can be operated selectively, the more comfort will be obtained from it. This means being able to start and stop the warming of any room on the system. The only clearance needed by ducts passing through walls is that which is provided by the insulation.

There are three types of heater suited to stub duct installations. They are:

1. A gas-fired air heater, in principle like that shown in *Figure 10.3*.

2. An indirect heater, as in *Figure 10.4*.

3. An electric heater of the type called Electricaire, which is essentially a large version of a storage heater, with incorporated fan and duct spigots.

If we compare these, the first could be the most awkward since it makes conditions. It must have access to a conventional flue or chimney, or else it must back against an outside wall so as to have a balanced flue. Combustion must, of course, be permitted. Since the other two do not have these restrictive conditions, the first survives mainly because it can incorporate a water heater if required, thus economising on a second flue of either type.

The second, having a boiler somewhere in the installation, carries its own solution to the domestic hot water question.

Only the third does not have any direct connection with hot water, and its great virtue is that it needs neither gas pipes nor water pipes run to the site – nothing but a substantial electric cable. But being a storage unit it does call for a strong and perhaps strengthened floor. If in doubt it might be best to bring a solid base up through a suspended floor, and frame the floor around the island site.

It will be seen, however, that the three kinds of heater each have a place according to the householder's circumstances.

The possibility of putting the heater in the cellar, in properties which have oné, should be considered, and it might be very convenient for floor registers. With a flued heater the nature and run of the flue has to be settled, and the gas people can very likely offer suggestions. A cellar

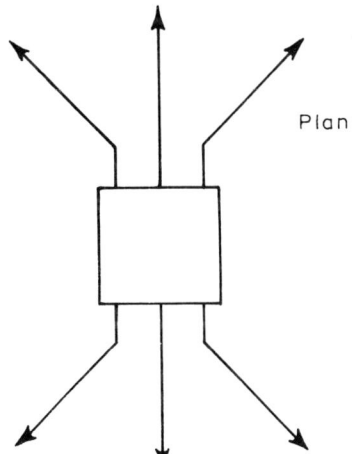

Plan

Figure 10.6. Radial stub duct run

having no other important uses could accommodate a radial duct system, as shown in *Figure 10.6*. No duct should be more than 6 m (20 ft) long, and it should have no more than two bends – not difficult conditions for most cellars.

It is often convenient to arrange for one discharge to take place in the hall, not only to provide a warm welcome but

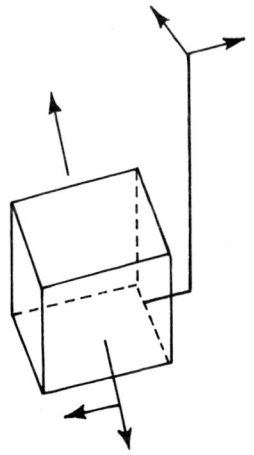

Figure 10.7. This heater includes a rising duct to warm a floor above

143

because in most buildings the warm air in a hall soon finds its way upstairs. It is, however, possible for a stub duct system to warm the first floor more directly. Using any shape of heater this can be done as shown in principle in *Figure 10.7.* Or you can get a form of heater known as storey-height in which the casing has an adjustable upper part, to disguise the extension of an open flue and also an internally-run rising duct. How such a heater can be made to serve at least three ground-floor rooms and at least two on the first floor is shown, in simplified form, in *Figure 10.8.*

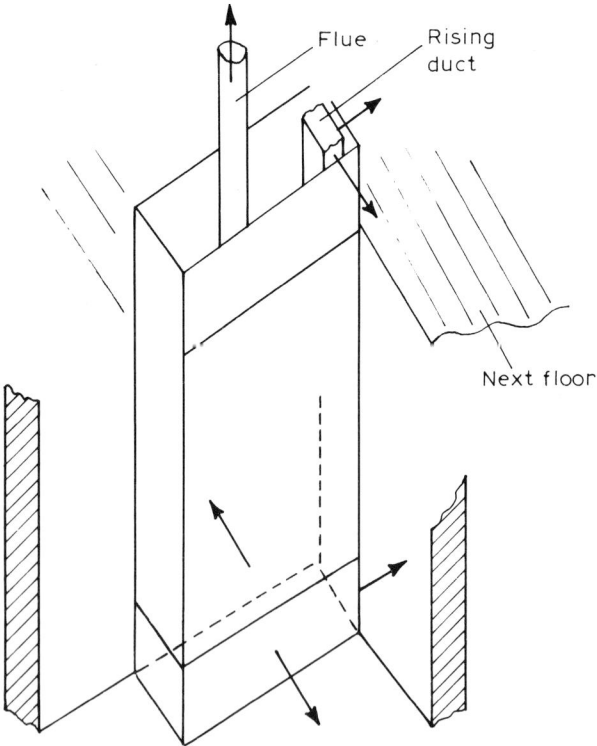

Figure 10.8. Storey height unit in a corner position feeding two floors

The same scheme lends itself well to a simple way of obtaining a return air duct. That is to take advantage of the duct-like shape of a channel between ceiling joists. As stated earlier, there is no compulsion that this shall be air-tight, nor insulated, and the most that need be done, if anything, is to line it with hardboard.

For all but the most low-pressure fanned systems it is usual to fit a balancing damper, known as a stackhead damper, prior to the register. Like its wet-system counterpart, this stackhead is adjusted to a fixed position and plays no part in the daily on/off operations. Although the air velocity issuing from a register should not exceed 2 m/second or about 400 ft/minute, every effort should be made to place registers where their discharge will not normally impinge upon people where they will be seated.

Radiant and convected heat

The basis of central heating, which is warmed air, is a mixed blessing, a convenience which is far from ideal. It has to be compared with radiant heating, where it scores the following points:

● It does not involve high temperature sources, which are a potential danger.

● It is pervasive, i.e. it spreads around corners and fills a room. Radiant heating, by contrast, travels only in straight lines. (A gas fire warms a room by secondary convection, i.e. it warms furniture, which gives off warm air currents.)

The main attraction of radiant heat lies in its beneficial effect, which may be said to work on the mind as well as the body. We feel more satisfied looking into a glowing fire, or enjoying the sun, than simply sitting in a warm room. Moreover, people in radiant warmth can tolerate quite a large drop in the ambient (air) temperature, and one experiment showed a fall to 8.5°C (47°F) before any discomfort was felt. And did we not, when young, play on icy fields stripped to the waist when the sun was bright?

145

There is a practical moral to this. One should take every opportunity to introduce a radiant source into the house, whatever the main heating system. All background heating systems rely upon local sources to make up, and quite often a gas fire in the living room serves as a booster of both temperature and morale, and also is ready for use in mid-season, when it is not worth lighting the boiler just for the evening.

11

Central-heating controls

The ideas of central heating and of control go together. Central heating stands for heating under control, and to a large extent automatic control. But like everything else control can be overdone, and we need to know what functions to control and how to go about it. The three factors which stand for comfort with economy are 'how much warmth', 'when to warm' and 'where to warm'. The object of controls is to serve these three factors, and to do so without overlap or undue extravagance in first cost. Not only is it wasteful to pay for aspects of control which you will not use, it is more important to note that two controls cannot work at the same function without contradicting each other.

Some controls are not optional, being supplied as a part of an apparatus. Most obvious is the boiler thermostat, or its counterpart on an air heater. Other supplied controls include those on a fan convector. But we are concerned with those which we must choose for ourselves.

How much warmth?

The answer to this is a thermostat, an instrument whose function is to maintain at a steady level a temperature which has been preselected. Chief among these is the room thermostat or roomstat (*Figure 11.1*). Some makes of roomstat have attracted to themselves secondary functions, such as:

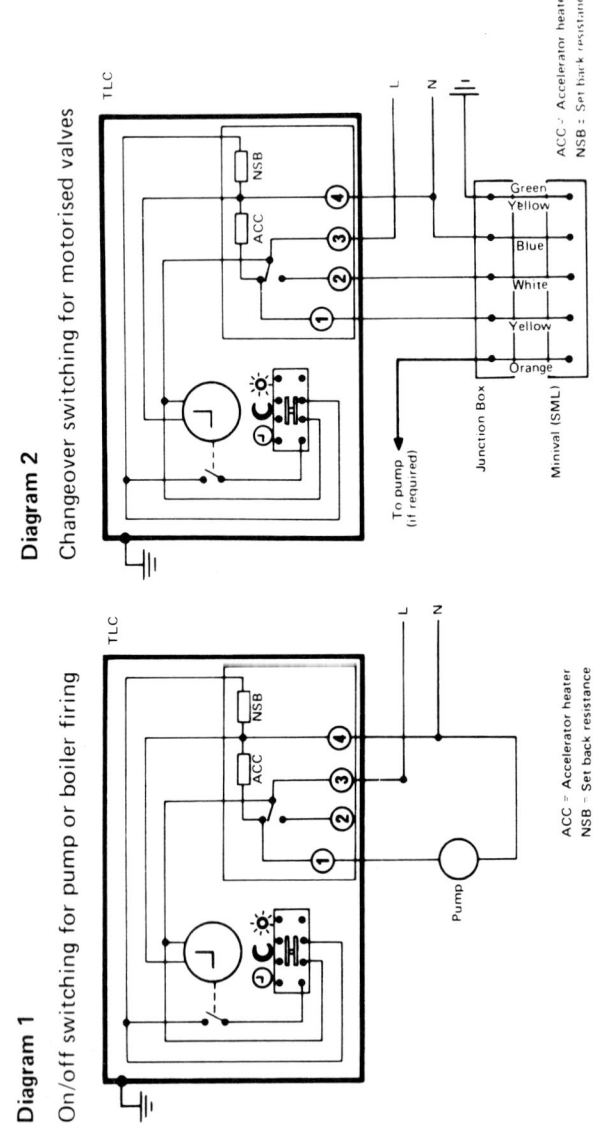

Diagram 1

On/off switching for pump or boiler firing

ACC = Accelerator heater
NSB = Set back resistance

Diagram 2

Changeover switching for motorised valves

ACC = Accelerator heater
NSB = Set back resistance

Figure 11.1. The Satchwell Sunvic Clock Thermostat TLC. Diagram 1 shows how both time and temperature may be used to switch on/off a pump or boiler; diagram 2 has the use of the same functions effecting a change-over between motorised valves. The three symbols are a three-way switch offering, left to right, automatic switching from normal to reduced temperature; override to continuous setback temperature; override to continuous normal temperature

- a simple switch to control the whole system on/off;
- a minimum temperature stop, allowing an easy flick back to a 'slumbering' temperature;
- a timed setback, by which the temperature may be set lower (to a predetermined temperature) for a period up to 12 hours, after which it returns to normal;
- clock-controlled setback (as above) with override switch;
- heater accelerator, an internal device which speeds up the response of the instrument to changes of temperature and so reduces the temperature swing. This is common to most roomstats.

There is no one answer to the question whether to choose a model with an extra feature. Many people, for instance, prefer to adopt a lower working temperature at night, instead of Off, and there is much to be said for it. If, however, you have no intention of working in that way, buying setback would be extravagant.

There are a few simple rules about installing roomstats, which are in sole charge of the house or, in a zoned system, of a zone. First, it must be placed in a room which will be warmed all the time the system is at work. That is why it is so often found in the hall, and no one must suppose that the hall temperature will dictate the living room temperature. The apportionment of warmth between rooms is governed entirely by the size of the heating equipment – radiators etc. – installed in the room. Thus a correctly-designed system with a roomstat in the hall set at 15°C (60°F) will bring the living room up to 21°C (70°F) or whatever is chosen.

The second need of a roomstat is to be open to the room but not too much exposed to it, like a spectator trying to keep out of the wind. The position to choose will be about chest- to shoulder-high, in the main body of the room where the roomstat can sense the true room atmosphere, but out of the way of cold draughts (from an opening door or window), through draughts from door to door, coat hanging, and stray influences such as sunlight in windows, heat rising from warm pipes or radiator, table lamp or television. When choosing a place, one should try to take account of present and future happenings in that area.

149

The roomstat, like most controls, acts simply as a switch and is wired back to the nerve centre of the system. You have a choice about what it will switch, whether it will be the circulating pump, or the whole system, i.e. boiler and pump. (We are assuming the system to be the average one, a boiler supplying heating and hot water.) These are the criteria.

If you switch off the pump only, the boiler stays at work to supply hot water. Also, during a period when the heating is in principle On, the boiler remains at the ready when the thermostat and pump call for hot water again. Pump switching is the usual method.

In the now-rare event that the system depends upon gravity circulation, switching off the boiler is the only possible choice, unless magnetic valves are used, and neither prospect has much to commend it.

There is one important exception. Solid-fuel boilers which include electrical parts, e.g. forced-draught fan, or chain stoker, can be switched off, but the rate of response is unacceptable. Switching the pump is therefore the only way, and it is *essential* to include a heat buffer (commonly an uncontrolled radiator) in the domestic hot water side of any circuit which includes a solid-fuel boiler and switched pump.

An important element in the choice of a roomstat is its electrical characteristics. It is unlikely that the permitted amperage will be exceeded in any domestic installation. But voltages may be mains or, sometimes, low at 6 V, 12 V or even 24 V. It is clearly necessary to be consistent with the system.

Radstats

A roomstat has various limitations, however. It is responsible for the entire house on a basis decided before the system was installed; in particular, it has to control the inflexible heat ratios in the system. Surely there must be days when even one room departs from the pattern, in particular remains unoccupied when occupation was expected. That is a waste of heat, and if you multiply that one room by the number of days per year the unexpected happens, the answer could be a lot of wasted money.

It follows that any way of achieving the purposes of a

150

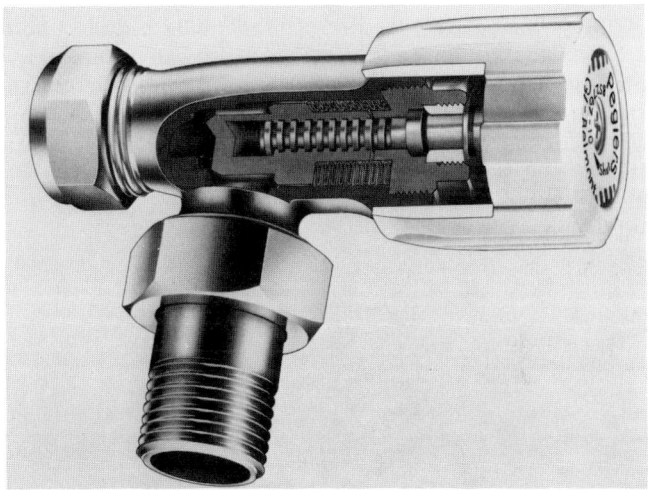

Figure 11.2. Pegler's Belmont GL glandless radiator valve, non-thermostatic. The expansion bellows creates the leakproof seal. Suitable for water to 120°C and 10 bar pressure. This valve is used for manual control of individual radiators in a system controlled thermostatically by a roomstat; also to effect on/off control of other radiators in a room where the master radiator is thermostatically controlled by a radstat

roomstat without incurring the wastage mentioned, even though it might cost more to install, will eventually pay for itself. That is where radstats – radiator thermostats – come in. They represent individual control of radiators or, if there are two radiators, of individual rooms (*Figure 11.2*). As well as being a ready means for shutting off unwanted radiators and putting them to work when wanted, they are able to make further savings. The broad control of a roomstat can result in individual rooms being too warm or too cool *for the occupants at that time.* (Chapter 12 has some notes on psychological and physiological influences on reaction to ambient temperatures.) A ready means of altering the room temperature can therefore affect the overall economy of running, and overheating is probably more common due to numbers of people present, and gains from lighting, TV and so on. A number of reputable firms now make radstats.

This is a non-electric control, depending usually upon progressive throttling of the water flow by a valve actuated by a thermostatic bellows. Most models now include a means of locking the set temperature and/or limiting the range by which it can be varied by casual interference. The Drayton TRV2 is one of those which modifies the Off position so that it will open automatically should the ambient temperature fall dangerously low.

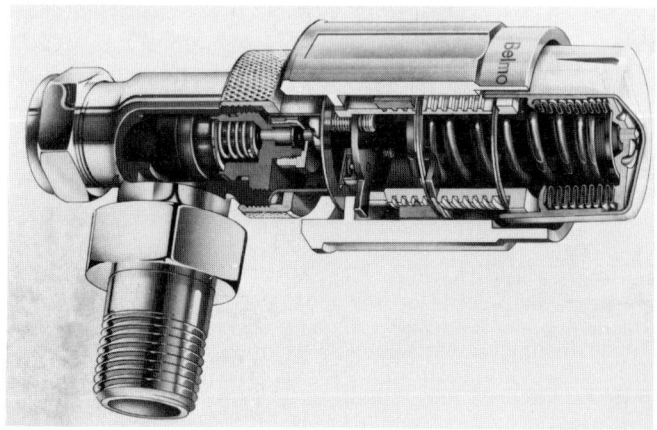

Figure 11.3. Pegler's Belmont Thermostatic radiator valve. Calibration range from 10°C, but the handwheel may be locked in any position. Maximum static pressure 8.27 bar

Danfoss, one of the longest-established makers, offers the self-contained unit and also one with a remote sensor, so that sensing can be done in a more neutral zone than adjacent to the radiator. Danfoss also makes a remote adjuster, in case the radstat happens to be difficult to reach. That such luxuries are not universally used is doubtless because they have to be paid for, and it will take a little time for them to become the norm.

At this point, let us clear up two common misunderstandings. The first could arise from advertising campaigns carried out by, or on behalf of, controls manufacturers. Rightly stressing the importance of controls in achieving economical and comfortable heating, they could lead anyone to believe that economy is inevitable once enough instruments are in place. That is like saying that if you buy a Rolls you are guaranteed against road accidents. Economy, like safe driving, depends on your handling of the equipment, not on the equipment alone.

The second point bears upon this, if you are led to believe that with controls it is 'the more the merrier'. You must not have more than one control doing one job, otherwise you get confusion, and what is called 'hunting', the controls vainly trying to catch up with each other. That is why we have read with some surprise that installations equipped with radstats should have also a roomstat. It is likely that the reason behind this is a wish to avoid a continuously running pump, which is needed if each room is autonomous. In those conditions only the clock would stop and start the pump. Let us make it clear that if a roomstat is required, for that or any other reason, it *must* be set to a temperature significantly above the highest temperature called for by any radstat. And bearing in mind that, as pointed out, 60°F in the hall might be equal to 70°F in the living room, this must be done finally by experiment.

When to warm

We now turn to clocks and time control. This is an essential ingredient in the control pattern, for you want events like stopping and starting to take place by pre-arrangement. If you have to get up on a cold morning to start the heating system, what is a heating system for? Equally, you could forget to turn it off at night. So comfort and economy as well as maximum convenience depend upon time control.

Clocks
Horstmann used to make a hand-wound clock which had the ability to open and shut a gas cock. Modern clocks are either electric or electronic, which of course presents endless

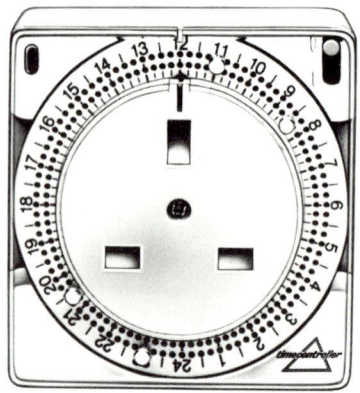

Figure 11.4. The Time Controller by Smiths Industries in either 24-hour or 7-day version. This is the all-purpose clock, which in 24-hour mode is capable of 24 cycles per 24 hours. It may be used with such appliances as oil-filled radiators and, like them, is portable. It simply plugs into a 13-amp socket

opportunities for variation, addition of functions and integration of functions. *Figures 11.4, 11.5* and *11.6* show the wide variety of forms taken by devices for time control. Let us begin by thinking of the basic clock. You do not simply buy 'a clock controller', or time switch. It must be capable of handling the amount of current which, for instance in direct switching of electric heating, could be quite large. In such cases you should always take advice from the seller. But we

Figure 11.5. The Sangamo Model 610 clock or time-switch, with 24-hour dial and available with two pairs (on and off) of tappets, or three or four pairs – reducible to one pair by removal of the unwanted ones

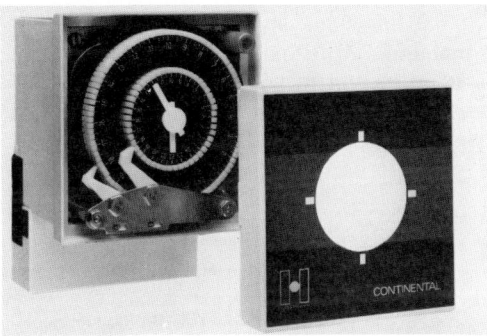

Figure 11.6. One of the versatile Horstmann Continental range of time controls. It has two 24-hour dials, the outer with 96 tappets, the inner with 48. Performance can be varied from one operational change per week to 48 changes per hour. There is battery back-up in case of power failure

can see some of the pitfalls simply by thinking about a straightforward hot water system.

What do you want the clock to do and how do you want it done? For instance, the clock could control the circulating pump, which means the heating only; or it could control the boiler, which means both heating and domestic hot water. In the first case the controlled item is likely to be at mains voltage. In the second case, the boiler controls might be at mains voltage but are more likely to be at either 6 V, 12 V or 24 V: in short, low voltage. Now follow the L wire in *Figure 11.7a*. It goes to the clock motor, on to the switch gear and then to load. The N goes to clock and load. This is called a three-wire clock, since of the four possible wires (L and N to motor, L and N to load), N is common to both circuits. Obviously the entire wiring is at mains voltage, and so the load must be working at mains.

Compare this with *Figure 11.7b*, where you will see that the pair of wires labelled Motor circuit, which are usually at mains voltage, serve only the clock motor. The other pair, labelled Switch circuit, do not interconnect at all with the first

155

pair, but are switched mechanically and pass to load. Absence of interconnection means that the four-wire clock can be used for any combination of voltages (subject to maximum amperage of course). It is, for instance, the correct choice for fitting into a low-voltage job such as boiler control.

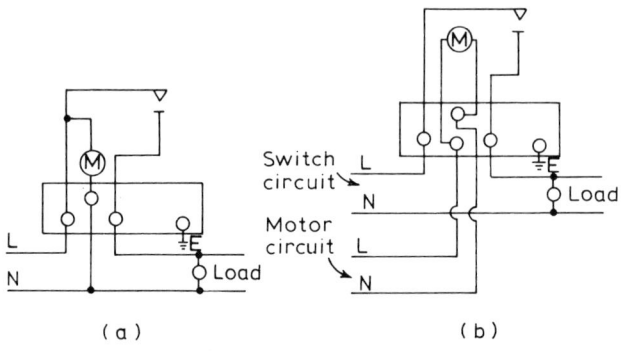

Figure 11.7. (a) is a diagram for a three-wire clock, for use when the clock motor (M) and the switched load have a common voltage. (b) is a diagram for a four-wire clock suitable for equal voltages or wholly dissimilar ones, usually high and low

In short, then, electrical suitability is the first criterion. Next we must decide what sort of performance we want, from the increasing variety available. Take the relatively simple Smith TS100 Time Controller. This can be programmed for up to 24 on/off cycles per 24 hours, principally because it has many other uses than the control of heating. Or look at another simple unit, the Sangamo S611. In its Form 2 it can make two on/off switchings per 24 hours. Form 3 and Form 4 provide three and four on/offs. But then, the omitting device can cause omission of either On or Off twice per 24 hours; and there are 'early on' and 'early off' facilities as well as 'late on' and 'late off' ones. The possible permutations on these are enormous. The omitting device, or override, is a simply-operated means of causing the status quo to be interrupted, e.g. if the clock is On it goes Off. Return may be automatic at next clock change, or it may in some cases have to be restored manually.

156

With so much choice you have to decide what you want. At the beginning of Chapter 12 there is a timetable to cover an average family working week. You will see that this envisaged only two On/Off periods in 24 hours. The weekend is not, probably, a routine occasion. If you are staying indoors, or if you will be away all day, the override mechanism is sufficient to take care of that, without altering clock tappet times. The four-tappet clock, i.e. 2 On and 2 Off, is by far the most popular because it generally suits people's behaviour.

It is possible to make a case for a six-tappet clock, one which has three On/Off changes per 24 hours. Suppose, for instance, that the household gets up and away very early, comes home at midday, and finally home in the evening. What is important is the knowledge that, however you assess your movements, there is a clock that will match them. If the controlled apparatus is self-contained and lightweight there can be no objection to arranging 10, 12 or more starts and stops in a day. But it would be unwise to treat a wet system and boiler in this way. In such cases where a heavy appliance is heated and left to cool, a considerable amount of heat is actually wasted simply in heating the appliance. If you multiply this waste by even four times in a day, the total wastage becomes very large. If more than three On/Off periods were required, it would be worth considering whether to choose instead a High/Low arrangement, since the warmth during the Low periods would not be wholly wasted but would be absorbed by furniture etc. and returned gently. But each case deserves to be considered on its merits, taking into account the durations of On and Off, and the nature of the use, whether for comfort or other purpose.

Programmers

The most prominent derivative of the clock is the programmer. In electrical models the clock trips cams which operate switches and by means of apparently complex wiring these switches cause functions to start or stop. We use the word 'apparently' deliberately, since detailed inspection of a typical wiring diagram shows that it is ingenious in concept but not difficult to follow.

157

Figure 11.8. The Horstmann Emerald time-controller for a single function for circuit e.g. boiler only, air heater, immersion heater, electric convector. It offers four programmes, viz continuous on, continuous off, on/off once per 24 hours, on/off twice per 24 hours

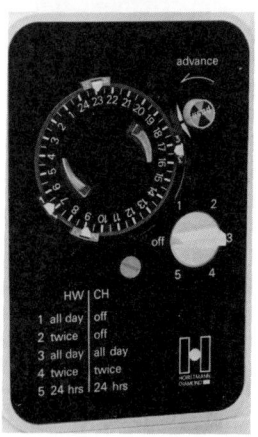

Figure 11.9. The Horstmann Diamond time-controller able to deal with a double circuit, usually hot water and heating. In addition to continuous off it offers a choice of the five programmes marked on the face

A typical programmer bears some resemblance to an automatic washing machine, in that it offers the user a number of programmes. Naturally the number of programmes and the cost of the controller rise together, making it desirable to understand what is being offered. There are, one supposes, people who regularly use every programme on their washing machine, and every one on a multi-programmed control on the heating. But we have not met any, and so feel safe in suggesting that for most people a small number is all that is worth paying for. We will look at what is programmed, and a typical set of programmes.

Let us start with what is still the most common form of installation, where the heating is pump-circulated, and the domestic hot water, under gravity circulation, depends only upon the boiler. The pump, and the boiler through its control system, can be independently switched. Hot water cannot be separately controlled, and clearly if the boiler is switched off to suit the hot water, the heating cannot be at work. But hot water can be On while heating is Off. From those simple facts we can construct a programme which is matched to the four-tappet clock, i.e. two On and two Off periods.

1. Continuously On, both, and subject only to roomstat and cylstat.
2. CH (central heating) once, HW (hot water) once a day. This means that only the first On tappet and the last Off tappet are being used.
3. CH twice, HW twice a day. All four tappets are being obeyed.
4. CH Off, HW once a day. Only the boiler is at work.
5. CH Off, HW twice a day. As for 4.
6. Continuously Off, both CH and HW.

That is typical six-programme routine, sufficient for most people. It can be added to in many ways, for instance in having HW on continuously, with CH either once or twice a day.

Even the programmer has grown in stature since electronics came into it. The Smith Centroller 2000, for instance, is described as electronic microprocessor-based (*Figure 11.10*). It is one of a breed that takes account of an advance in the system, namely pumped primaries for hot water*. This means

---

* Combined heating and hot water systems described in this book have pumped heating and gravity-circulated hot water systems. By pumping the hot water system, using a common pump or a separate pump, a greater amount of control can be exercised over the hot water side, e.g. by being able to run heating only and, as in this example, by having full control of hot water temperature. Detailed examined of pumped primaries is beyond the scope of this book.

Figure 11.10. The Smiths Industries Controller 2000 is a central heating and hot water programmer based upon an electronic microprocessor. It includes override, one-hour battery reserve for mains failure, automatic resetting after a power cut, and a number of possible instant changes of programme by push button

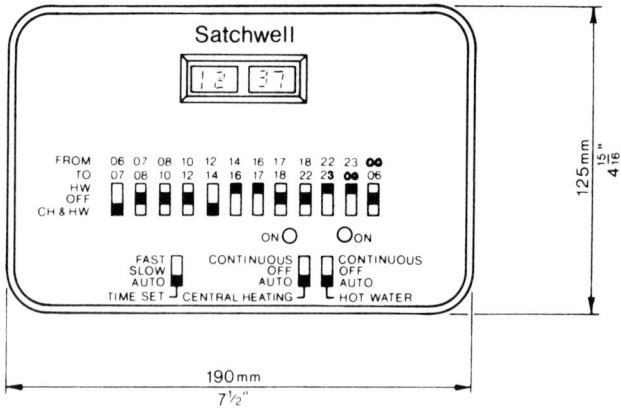

Figure 11.11. Satchwell Sunvic Electronic Programmer DHP 2201. This programmer gives in the simplest form detailed control over domestic hot water and central heating, including the ability to interrupt any part of a set programme. Note that finger control over 12 × 2 hour time-segments has replaced the system of tappets around a clock-face

160

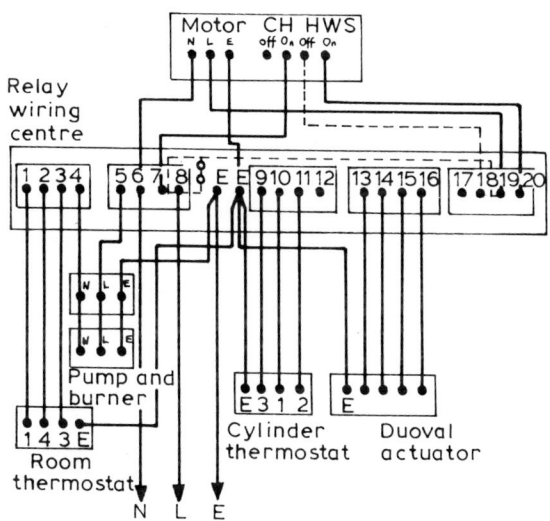

Figure 11.12. Typical installation wiring diagram for the Satchwell Sunvic Duoflow system. The Duoflow fits into systems which use a common pump for heating and hot water, and switches heat to either according to need as sensed by roomstat and cylstat. When both are satisfied it shuts off the boiler and pump

that CH and HW are capable of totally independent control. But there is more to the Centroller than that. It allows you to punch up four temperature levels for heating, and also to determine, on the controller, the temperature of the stored hot water. Clock display is digital, which is rather more convenient than trying to locate a figure on a 24-hour dial against a metal indicator.

At the extreme end of progress, at least for the present, are such devices as the AMF Paragon EC 403. In the interests of energy saving it will, for instance, receive instructions about what time a building should be up to temperature. It will then sense a number of contributory factors, outdoor temperature and weather, building temperature, and so on, and will decide for itself what time to start the heating. It is not yet on the domestic market.

Another microprocessor-based control system, the Chronex by L.E. Lloyd (Cambridge) Ltd, which is available for domestic use, seems to have uses one would not even imagine. Having a memory it can be programmed for years ahead. It will switch lights on and off, work a burglar alarm system, all in addition to keeping control of heating. Its makers say that for what it does and saves, the cost is modest. They claim too that programming it is similar to using a good-quality calculator.

There is a highly subjective element in making a choice. Modern youngsters brought up on computer games would probably cope very well with the more intricate controllers. But at one time we judged the acceptability of every bit of consumer goods from, as nearly as we could, the standpoint of a nervous elderly lady living alone. We should not lose sight of that criterion.

Where to warm

The answers to this, the third of the three conditions, are not as complex as the other two. Domestically, it will usually mean dealing in one room at a time, though there is an important exception to this, as we shall see. It would be altogether too complex and expensive to attempt to control the flow of water to each radiator from a central source. The only way to do that would be by fitting a solenoid valve and inviting more maintenance, in each room or at each radiator. But to have to do this would be quite unreasonable. We have not made any such demand in respect of electric light, being perfectly content to switch on and off as we come and go. The most we should ask, then, is that controlling the heating should not be much more difficult than that.

Such a condition does show up the common radiator valve, unless it has a 'fast' thread, and will go from open to shut in about half a turn. To bother about 2½ turns every time calls for great dedication to economy. That is where the radstat or thermostatic radiator valve scores another point. It does turn off, or down, very easily, and it will already have been accepted as a part of the control system simply by being able to control room temperature.

A different case allied to the wet system is that of the fan convector. This, being electrically operated, can be electrically controlled, by manual switch (just like the lighting) or it can be switched automatically, by a clock or thermostat. (It has an inbuilt thermostat which helps it to keep a stable room temperature.) Most have another manually selected feature, the ability to discharge warmth at full, half or minimum power.

Figure 11.13. Sangamo S 414 Twin-Set two-zone controller. It allows for time control of each zone independently and, as the wiring diagram *Figure 11.14* shows, the separated circuits can then have the extra control of a roomstat. In fully-pumped systems the S414 will give independent control over services, not only central heating and hot water

Room-by-room control, then, seems to lie for the present with the personal attention of the householder or user of the room. If we consider larger areas, such as one floor of a house, or it may be the front or back half of a house,

163

individual control can be left to automatics, and the principal items are called zone controllers. In normal circumstances a zone is an area which can be served by a subcircuit, with reasonable economy of pipework in order not to incur high installation cost. *Figure 8.10* (Chapter 8) illustrates the principle of zoned circuits, in that case zoned by floors. That is a perfectly logical division, since bedroom times and temperatures do not coincide with the requirements of the living rooms. More than two zones can be organised, and even subzones, so long as the entire system is within the capabilities of the power units, boiler and pump. Such complication

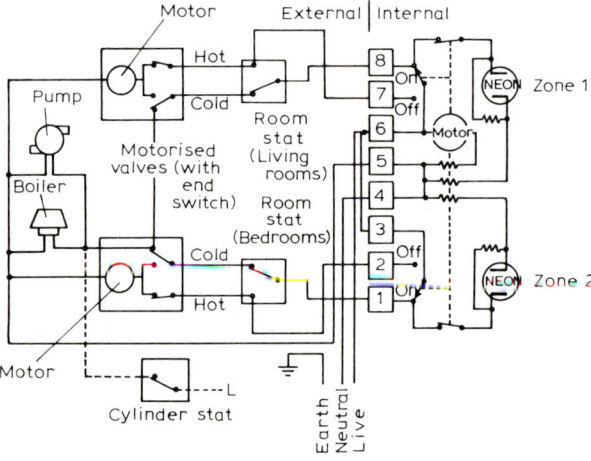

Figure 11.14. A typical wiring diagram showing the internal wiring of the Sangamo S414 Twin-Set controller, wired into two zoned circuits. Do not use this type of diagram without first checking that it suits a particular installation

is rarely required. More likely is a split into horizontal zoning, supposing that the back and front of a house were in some way different, in occupants or reason for occupation.

One of the earliest zone-control valves, still readily available, is the Satchwell Minival, described as an On-Off motorised twin-shoe valve for flow control in low-pressure systems. The Minival may be mounted in any position, which means that it need not occupy a prominent position. It can be controlled by any device which has a single-pole change-over switch, leaving the way clear for control by clock and roomstat. And this means that each zone may be pre-programmed for time and temperature and left to get on with it. One model has evolved from the original which has auxiliary switching. Thus, for instance, when the valve starts to close it can be made to perform other, usually but not necessarily related, duties. For instance, it could shut down the boiler, or the pump. But it could equally well switch off the light or the television.

The above notes are not tied to gas, or oil or solid fuel. Nor are they specific to wet systems. They can be applied to any kind of system which is capable of being switched electrically, with obvious modifications such as the absence of domestic hot water from a system of electric storage radiators. There is, however, one very important exception. A solid-fuel boiler is *not* directly controlled by electricity in the sense that if current is switched off it at once goes out. A solid-fuel boiler may depend upon an electrically operated fan or damper, but withdrawal of power from that item is only the first step in a quite lengthy process of winding down. During that winding-down process a great deal of extra heat will be generated which, if not given an outlet, would be sufficient to cause the boiler to blow up. The sudden stopping of the heating circulation pump could bring water flow to a stop too, and lead to disaster. For that reason *an escape route for surplus heat must always be provided.* It must be a route which cannot be shut, by accident or any other means, and the obvious one is via the domestic hot water primaries. The cylinder could act as a heat reservoir. But if it is up to temperature it will not encourage circulation. The correct step therefore is to fit a radiator or at worst a towel rail in the bathroom, without valves, and connected to the primaries. This point has been made elsewhere in this book, but we make no apology for repeating it.

The compensator

There is a form of control which is concerned with the amount of heat. It has some staunch supporters, among whom this writer has never been numbered, but it has to be mentioned as part of the whole picture. It is called, broadly, the compensator. It is most simply described in its original form. A normal design of wet system is constructed, with the single exception that a three-port valve is inserted in the flow from the boiler. To the three ports are connected (1) the flow out of the boiler; (2) the flow going on to the circuit; (3) a bypass leading up from the return pipe just before it goes back to the boiler. By this arrangement, some cooler water can be mixed with the flow, to modify the temperature of the water going forward to radiators etc. The reason for such modification is to vary the heat output from the system to suit varying conditions. What these variations are it is the duty of a sensor to report as signals to the three-port valve. The sensor is situated outdoors, but not in direct sunlight or direct wind. If the weather turns colder the sensor passes a message to the three-port valve, which closes against the bypass and increases the heat output from the system. And vice versa.

For years the flaw in this method has been house insulation, which has accelerated in recent times. The system remains the same, but the delay in outdoor conditions affecting indoors lengthens, until it can be as much as half a day before it becomes effective. But the delay in implementing instructions from the sensor is measured in less than an hour, which could lead to a long period of sweltering, or shivering. In the interval, supposing that the weather should change the other way, the whole system will quickly get out of step by being one or more changes behind the facts.

It is fair to add that the most recent makers in this field have introduced compensators to iron out the flaws mentioned, and no doubt these are effective. Anyone interested should examine critically any proposal to make sure that the above objections are overcome and that in fact the cost of such devices is likely to be justified by results.

166

Stop valves

It is symptomatic of the modern attitude to heating that we have not until now mentioned the oldest form of control of all: the common stop valve. Yet we continue to rely upon it at every turn. For instance every radiator, hot water convector etc should be equipped with two stop valves. Usually one is used to 'balance' the system, by admitting just as much water as is the share of its appliance, the other being an on/off valve. But in the event of a radiator having to be removed from the circuit for repair, or for painting the wall, isolating it by means of the stop valves is far better than having to drain the circuit. By similar reasoning, every circulating pump should have two valves, on inlet and outlet. The outlet valve could have a second and important function. Not all pumps are equipped with integral flow controllers, but control of flow might be called for. In that case it can be done by throttling on the *outlet* (not inlet) valve.

Equally there are some places which must never have valves fitted, for instance, the flow and return pipes to the hot water cylinder. And no vent pipe should ever be valved. The point is that if valves are fitted, someone might shut them, and in the cases mentioned a dangerous condition could soon follow.

We do not look enthusiastically upon dual systems, often home-designed, where the objective is to couple into a common water system the outputs from two different boilers which will be at work at different times. Stop valves are sometimes used, to isolate that item not at work at the time. But sooner or later someone will make a mistake, and the result could literally be disastrous. By the same reasoning, we have no sympathy with the enthusiast who puts into practice his own idea of a complicated circuit, valved in most cases. He undoubtedly operates it perfectly. But come the day his wife or child, or a new owner of the house, is obliged to try, and once more the ingredients for disaster are there. It simply is not worth it.

12

Comfort with economy

The most important principle involved in combining comfort with economy is stated in the three-point rule:

1. Have *as much* warmth as you need, and no more.
2. Have warmth *when* you want it, not at any other time.
3. Have warmth *where* you want it, not in any other place.

The implementation of the second and third rules can be left to instrumentation if your habits are regular. If, for instance, you always move into the living room on winter evenings at around seven o'clock, you can get a clock to change the heating over from one area to another, provided that you have a zoned system. If you do not, then it is no great hardship to earn your economy manually. It means putting the living room heating to work by about 6.30 p.m., opening a radiator valve, or switching on a fan convector, or lighting a gas-fired wall heater or a simple gas fire: then, at 7 p.m., shutting off whatever warmth was in action in the room you are leaving. Maybe it sounds troublesome, but you do it with the lighting.

Those comments relate 'where' and 'when' to individual rooms. But 'when' is also very relevant in the major context of the house itself. You can fit your own details to a sample weekday programme which looks like this:

7.00 a.m. Heating starts work.
7.30 a.m. Family gets up.
8.30 a.m. Last of family leaves home. Heating goes off.

4.30 p.m. Heating starts work.

5.00 p.m. First of family arrives home.

11.00 p.m. Last of family goes to bed. Heating goes off.

The allocation of warmth to various rooms takes place *within* that broad programme, which is the concern of the primary unit, the boiler, air heater etc. Being so fundamental it can be left in the hands of a clock or programmer, the programmer including a clock. The programme shown requires what is called a four-tappet clock, able to give two On and two Off periods in each 24 hours. At weekends and other

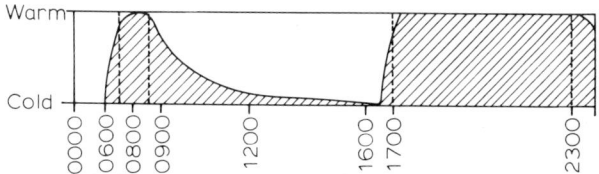

Figure 12.1. A graphical illustration of the actions of heating and cooling between 6.00 a.m. and midnight. Note that whereas heating is power-driven and the curve rises steeply, the rate of cooling is proportional to the excess of temperature over the surroundings, and is therefore much slower. Because of this, the heating may be shut off 15 minutes before it is no longer wanted, and this can save more than 5 per cent of the total fuel bill

periods when the routine is varied the clock can be reprogrammed, or more usually countermanded by the override mechanism, usually a simple lever. Or, at greater cost, you can buy a clock which has more functions, including the ability to deal with odd days, or even with each day as a separate entity. More was said on this subject when discussing controls, in the previous chapter.

Comfort

The first rule seems straightforward. A person feels comfortable in a temperature which depends almost entirely upon his physical characteristics at the time. It is a choice dependent upon age, health, tiredness, among other factors, and also a predilection for cooler or warmer ambience. This shows what nonsense it is to be dogmatic about a point on a temperature scale, so we can do no more than indicate maximum and minimum limits, for example 21–23°C (70–73°F) as suiting the majority. The rule is simply saying that you should feel neither chilly nor over-warmed, and that is the key to comfort with economy so far as the existing installation goes.

But it is not the entire answer, by any means. The person experiences chilliness because of loss of warmth and depends upon his surroundings to make up the loss. The surroundings also suffer loss and rely upon the heating system. And if the *rate of loss* in the second case can be reduced, a satisfactory temperature will be maintained for longer, and the person will enjoy the warmth for the use of less heating. Rate of heat loss, as we saw much earlier, is entirely to do with the various forms of insulation.

It is therefore adequate efficient insulation which is the key to keeping the first of the three rules.

The nature of the insulation can have its effect upon the other rules too. Let us take the important case of wall insulation, which may be done in three ways: on the outside, in the cavity, or inside. Each method has its pros and cons, dealt with elsewhere. What concerns us here is what happens inside the warmed house.

Let us look first at exterior insulation. Warmth is put into the house by the heating system and the warmed air warms among other things the walls. These, being heavy, will absorb a lot of warmth, right up to the barrier of insulation. The fact of that absorbing means that general warming, up to a given temperature, will take longer than if the walls did not count. And since, in spite of insulation, some cooling will occur when the heating is shut off, every rewarming will take proportionately as long.

Then consider cavity insulation. We can eliminate the influence of the cavity and of the outer leaf of brick, and so roughly halve the amount of heat absorption, and hence time taken to come to temperature.

Lastly we come to indoor insulation, or thermal inner lining. Subject only to the efficiency of the insulator, this leaves no brickwork to be warmed, and consequently warm-up time can be very fast. This has been proved in practice.

Allowing that that is not the entire picture, you can see how it will influence the lead time allowed, the time gap between putting heating to work in a room and the time of room occupation. Going back to our tentative daily programme, you will now see how the half hour allowed for warm-up could be either too short or too long depending upon the insulation details. And so you take account of such details when making your own programme.

The point is sometimes debated whether it is more economical to shut off heating overnight, or merely turn it back and so avoid fully reheating next morning. It is precisely the same argument about immersion heaters, whether to shut off or leave on.

There can be no doubt about the answer in both cases. It is undoubtedly more *convenient* to let systems run on, but unless you can cost convenience it must be less *economical*. The simple reason is that heat loss is proportional to difference in temperature, and if indoors is kept warm (or the cylinder kept hot) there is an appreciable loss of heat over some hours.

If heating is off then the system cools and heat loss is less. But if the insulation were perfect, convenience would win since there would be no difference. And so the better we make our insulation, the less important does the difference become. Since in the case of solid-fuel appliances we have no choice but to leave them on, though 'slumbering', there is even more incentive to put in first-class insulation.

In the case of electric storage heaters, some current models have gone back to the floor-heating days and include automatic input control. This enables the user to make a judgement about the next day's weather, in relation to today's weather and input setting. There is some scope for economy here, as there is on the output side. This can be controlled to make sure that the heater output is boosted only when you require it, whether in response to an automatic control (clock or roomstat) or manually or by overriding the control.

Perhaps the best general advice to the user is: 'be flexible'. Controls, even on/off valves, are there to be used. A room temperature controlled by a roomstat in the hall might be well suited to two occupants. But bring in six friends and in no time you might usefully close down a radiator valve, turn back the fire, or close the damper on a solid-fuel fire. Or of course with selective heating, turning back the roomstat might be the simple answer. But the principle is the fundamental one of not having more warmth than is necessary.

We are only too well aware that one cannot be too dogmatic about how a system should be run, and we have elsewhere made the point that heating, like the Sabbath, is made for man. How for instance can one possibly justify keeping heating on in a house which is empty for weeks? Well, keeping freezing at bay is one good reason. A second is to keep condensation out of fabrics and furnishings and so perhaps save much more money than is spent on the heating. In such cases, and many more like them, the only stipulation to be made is that the minimum use shall be made of energy. And if we look back at *Figure 12.1* and the slow cooling curve it will be seen that a little heating time can go a long way. In practice, for instance, a reasonable frost repellent in reasonably-insulated premises need be no more than two to three hours soon after midday and soon after midnight.

The practice which really does call for censure is wallowing in heat, for which there are several possible reasons but no excuse. Lack of controls, ignorance, inertia, 'we can afford it', 'the landlord is paying' all appear regularly. People have been saying for a long time that every degree of temperature above what is needed wastes 10, 7 or 5 per cent of the total fuel bill and the fact of wastage is real, and of course unnecessary.

The object of this book has been to create an interest in the principles so that readers will reach informed conclusions about their own circumstances and form their own targets for temperatures, times and places of operation, and so on. We can, to quote a nineteenth-century divine, more usefully point the way than lead it.

13

Other sources of energy

Heat pump

This is a remarkable device in that it uses a certain amount of gas or electricity in order to produce a good deal more energy. Remarkable, too, is the variety of places from which it can draw heat: the soil, a pond, the air, and an ingenious combination, the rather weak output from a solar heating system in the UK winter.

The heat pump may be seen as a refrigerator in reverse. Instead of taking heat from its own interior and scattering it to the outside, it concentrates outside heat within itself. Domestically it is usually put to work heating water. Since all heat pumps start by using electricity or gas, the best place to start enquiring is at the appropriate showroom, since you can then rely upon hearing only about those which have received at least a measure of scrutiny and approval. And you can ask questions about application.

Renewable sources

Coal, gas and oil are not renewable, which is of course a present source of anxiety. But taking a wider view of energy, which includes the sun, wind, tides, and the forces which cause earthquakes and hot springs, many sources *are* renewable. Like the sun, what you take will not reduce its potential. All these sources are being investigated, but in the main they can be harnessed only at huge cost, and any success will mean electricity into the grid. They are not domestic matters.

There is much about the alternative sources to fire the imagination. The proposed Severn Barrage, huge offshore windmills, hydro-electric schemes, create great interest and if it were not for the huge costs involved, much work would no doubt have started. But we will not resist the temptation to draw your attention to the less glamorous truth, that nearly all those schemes are not essential. Several studies have shown that if every house in Britain was insulated to a high standard there would be such saving of fuel as would stave off any energy crisis for the foreseeable future. And the cost would be less than any of the more spectacular measures. So far as one can see, the biggest stumbling block to such a solution is that it is wholly unappealing to the imagination of ordinary people and of those in government. But once you become aware of the facts, you can do most of it for yourself, and in that way serve yourself and the country in the best possible manner.

Wind power cannot be said to be wholly domestic since it would be quite out of the question for every house in a street to have its own windmill. Only in the country, where there is space, can a windmill be accommodated, and it is therefore a minority matter.

Solar heating

The capturing of solar heat is another matter entirely, its effectiveness governed only by the supply. This means, in the UK, that it is at its best in high summer, and therefore excellent for warming a swimming pool. In winter, however, it can scarcely provide our domestic hot water, with hardly ever a scrap left over for space heating. But please note the reference to it under 'Heat pumps'.

Solar heating is conveniently divided into 'passive' and 'active', according to whether the apparatus has any moving parts.

In passive heating the apparatus may be the house itself. Or an interior, warmed by the sun through a window. There is a natural phenomenon called the 'greenhouse effect' whereby more heat can come in through sunlight than can get out via the window. This is because the sun's radiation is short-wave, able to pass through glass. But objects warmed by the sun to a relatively low temperature give off longer wave-length radiation which cannot pass through glass. Many ideas stem from this, including helping the process by having an automatic blind which opens when the sun shines, closes when it goes in or down.

It gives rise to such practical ideas as that a lean-to greenhouse is thermally better than a free-standing one since the back wall is a heat reservoir. Or, taking that a stage further, a lean-to greenhouse with its back wall made of water filled jerrycans, better able to absorb heat than a brick wall.

There are now many books on all aspects of solar heating, all worth reading though many Americans tend to regard the world as centring on California. British authors are preferred for British conditions. Anyone interested should make a point of reading two or more such books, not only to get a spread of ideas, but to assess the practical possibilities. There is nothing that a competent amateur cannot do in this respect, and the purpose of wide reading is to get enough ideas to make a good job. For instance, now that automatic controls have been standardised to take care of freezing, premature circulation of unwarmed water, prevention of boiling and similar events, attention is turning to storage.

There are, after all, drawbacks to energy which, almost like electricity, must be used while it is there. And even electricity is better since it can be conjured up at any time of day or night. Hence we find a man who built a cellar and filled it with stones in water, and it stores heat for weeks. On a more modest scale the hunt is on, and worth following. Another advantage of knowing something about the subject is that, if you decide to contract the job out, you are not at the mercy of any contractor with possibly outlandish ideas.

All we show here (*Figure 13.1*) is the simple construction of a solar roof panel. By connecting the water pipes to a pump the whole system is under control by instruments, details which the books make clear.

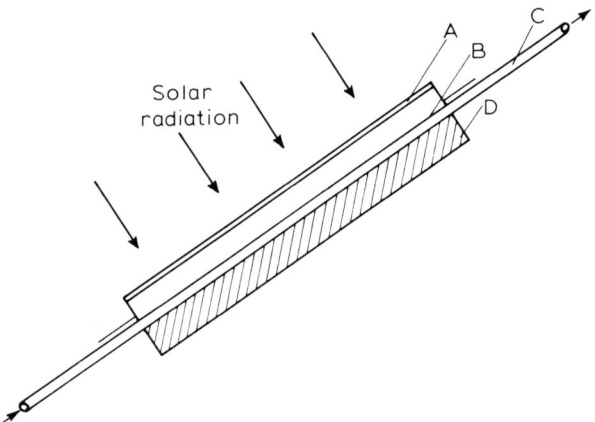

Figure 13.1. The construction of a simple solar panel, for roof fitting or free-standing. (A) Glass cover. (B) Black absorbent panel. (C) Water piping attached to B. (D) Insulation

A method of harnessing sunlight which has not yet arrived in a form which makes it domestically obtainable is the direct generation of electricity. The apparatus is called a photo-voltaic cell, used in large multiples, and it would be fair to say that development is proceeding well. It is worth looking out for in a year or two.

Biomass
A quite different but renewable energy source is called biomass. It may be described as burning vegetable matter to produce electricity, and is relevant to places where there is a surplus of vegetation; or where, as with the water hyacinth, an irrepressible growth may be cultivated, harvested and burned. But anyone who burns some of his garden refuse on an open fire is practicing the biomass system, though that it as near as it is ever likely to get to domestic application.

Reading and information

On solar energy, the following publications are recom-mended:

Practical Solar Heating, Kevin McCartney, Prism Press, 1978.
Solar Heating Systems for the UK, Wozniak, HMSO.
Solar Heating for Domestic Hot water – a guide to good practice. HVCA, Esca House, 34 Palace Court, London W2 4JG.

Information about solar heating and other forms of harnessing and conserving renewable source energy may be had from The Centre for Alternative Technology, Machynlleth, Wales. Send a stamped addressed envelope, or better still, pay a visit.

Heat pumps. If you cannot get the answers you want at gas or electricity showrooms, try the Heat Pump Manufacturers Association, % HVCA, Esca House, 34 Palace Court, London W2 4JG.

14

Where to turn for help

The practice of insulation is relatively free from compulsion via legislation. If you make a mess of it, that is your own affair. Probably the only area likely to concern the authorities would be fire risk, if you were to use some material likely to be a hazard. And even that might not become apparent unless there were a fire and your insurer refused to pay out.

One possible comfort to emerge from this is that you may ask for advice without sparking off an investigation by Authority. (We must exclude from that the case of grant-aided roof insulation, for which an inspection will most likely be made.)

Even the dedicated handyman must have his cavity filled professionally, and he should use a member of the trade association NCIA, the National Cavity Insulation Association, PO Box 12, Haslemere, Surrey GU27 3AN. This association protects the technical and ethical standards in customers' interests. At the same address are the National Association of Loft Insulation Contractors, and the External Wall Insulation Association.

These are contractors' trade associations, not information bureaux. There is an excellent source of information in Eurisol UK, the trade association of makers of mineral-fibre insulation. They are also very active in investigative work. Their address is St Paul's House, Edison Road, Bromley, Kent BR2 0EP.

You can lean upon manufacturers, and in particular those who lack a central voice similar to Eurisol UK. For instance, if you have any special queries about draughtstripping and your dealer cannot help you, ask him for the particular manufacturer's address.

You will find, too, that gas and electricity showrooms can often refer specific queries to a member of the technical staff who deals with insulation.

In contrast to insulation, practically everything you do with heating is the subject of some sort of control or regulation. The installation itself, the fuels gas and oil, electricity, and water are worth considering here.

The installation if it concerns the flue or chimney if fabricated or altered must be inspected by the Building Department at the town hall, and it saves a lot of trouble to consult them before taking any action.

Gas. The Gas Safety Regulations sum up one's obligations so far as the running and use of gas are concerned. If there is any mishap, which could easily amount to death or property destruction, the householder is personally responsible, whoever did the work. This means that a final check by the experts is desirable, for safety and efficiency. A call at the gas showroom will arrange this. Or you might engage a CORGI-registered installer. The setting-up of a boiler or other major appliance will be done by the manufacturer, for a fee, if required.

If all you need is advice, you can go to the showroom and ask to be put in touch with the appropriate person. Or, for the major appliance, you can contact the manufacturer, direct.

Oil. The town hall will have an interest in the siting of an oil-fired boiler from the point of view of fire risk, and they will probably insist upon the fitting of a fire valve. Save time by asking first. Also consult them about your idea of where to put the storage tank.

For advice on particular aspects of construction and operation, get in touch with the maker of the boiler. There is an approval authority, DOBETA, but its main function is simply approval, not public information.

Electricity. Few modern appliances work without electricity, and the user when installing has a duty to observe the

IEE Regulations, failing which the undertaker has the power to refuse a supply.

Sometimes the IEE Regulations make sense to the amateur, and can be seen at the local library. But for advice on particular points we always find that the local Electricity Board will respond to a hypothetical question, in the first place. This enables you to decide whether (a) you can do it yourself, (b) you need to engage a contractor, and (c) to ask the Electricity Board to attend to it for you. If they are alerted as for instance by coming to change over to off-peak, they are entitled to check the whole system.

Water. No wet system can be installed without some alteration to the indoor water system. Under the Water Regulations you are expected to give seven clear days (10 in Scotland) notice of your intentions, so that the Authority may inspect and pass the alteration proposed. That rule is a first class let out, for you will be told whether you are likely to contravene any of the other 54 Regulations. But if you want advance notice yourself you are entitled to see a copy of the Regulations at any time.

Other sources

There are other sources of help over which the threat of regulations does not hang.

For anything to do with solid fuel, the Solid Fuel Advisory Service (SFAS) exists to help you. If you cannot find a local office in your phone book, start at NCB headquarters, Hobart House, Grosvenor Place, London SW1X 7AE. SFAS are very good on chimneys, where site inspection is often necessary.

Building Centre. The principal one is in Store St, off Tottenham Court Road, London WC1E 7BT. One may get ideas from exhibits, advice from manned stations, endless information on products from the desk.

After long, sad experience we must warn readers that there is no substitute for professional advice, certainly not that advice which is so freely given in pubs and clubs by amateurs, however enthusiastic.

If you want to read up a subject, let us say solar heating or heat pumps, never underestimate the local library. Never mind whether you know any titles. It often seems that library staffs welcome a challenge, and they have many ways of carrying out a search, as well as getting a copy of any book.

If you are dissatisfied with the service from your local gas or electricity authority, each has a Consumer Council, and you can get the address of your regional one by asking at the showroom. The way to complain about the coal industry is though the Domestic Coal Consumers' Council at Dean Bradley House, Horseferry Road, London SW1.

Perhaps we may remind you too of your rights under consumer protection law, relating to appliances and materials. In all cases of faulty goods and materials you are entitled to redress from the seller. No longer must you put up with being told to get in touch with the manufacturer, for that has become the seller's job. If problems arise over that, go straight round to the consumer protection officer at the local town hall and let him sort it out. But remember that complaints are limited to faults in manufacture. If you have been experimenting and have caused damage you cannot hold anyone else to blame, and it does reinforce the message that expensive equipment is involved. It does deserve competent handling, for your own sake.

Index

183